Natural Eloquence

SCIENCE AND LITERATURE
A series edited by George Levine

Natural Eloquence

Women Reinscribe Science

Edited by
Barbara T. Gates and Ann B. Shteir

THE UNIVERSITY OF WISCONSIN PRESS

The University of Wisconsin Press
114 North Murray Street
Madison, Wisconsin 53715

3 Henrietta Street
London WC2E 8LU, England

5 4 3 2 1

Printed in the United States of America

Library of Congress Cataloging-in-Publication Data
Natural eloquence : women reinscribe science /
edited by Barbara T. Gates and Ann B. Shteir.
296 pp. cm. — (Science and literature)
Includes bibliographical references and index.
ISBN 0-299-15480-7 (cloth: alk. paper).
ISBN 0-299-15484-X (pbk: alk paper).
1. Women in science. 2. Science news. I. Gates, Barbara T., 1936– . II. Shteir, Ann B., 1941– . III. Series.
Q130.S39 1997
306.4′5′082—dc20 96-43668

Contents

Illustrations vii
Acknowledgments ix
Contributors xi

Part 1. Charting the Tradition

1. Introduction: Charting the Tradition 3
Barbara T. Gates and Ann B. Shteir

Part 2. Recuperating the Women

2. The Invisible Woman 27
Stephen Jay Gould

Part 3. Disseminating Knowledge: England

3. Fictionality, Demonstration, and a Forum for Popular Science: Jane Marcet's *Conversations on Chemistry* 43
Greg Myers

4. Constructing Victorian Heavens: Agnes Clerke and the "New Astronomy" 61
Bernard Lightman

Part 4. Disseminating Knowledge: Canada, Australia, and America

5. Science in Canada's Backwoods: Catharine Parr Traill 79
Marianne Gosztonyi Ainley

6. The "Very Poetry of Frogs": Louisa Anne Meredith in Australia 98
Judith Johnston

7. "Through Books to Nature": Anna Botsford Comstock and the Nature Study Movement 116
Pamela M. Henson

Part 5. Defining and Redefining Knowledge: Post-Darwinian Women

8. Revising the Descent of Woman: Eliza Burt Gamble 147
Rosemary Jann

9. Revisioning Darwin with Sympathy: Arabella Buckley 164
Barbara T. Gates

Part 6. Defining and Redefining Knowledge: Into the Twentieth Century

10. Conflicting Scientific Feminisms: Charlotte Haldane and Naomi Mitchison 179
Susan Squier

11. Rachel Carson and Her Legacy 196
Rebecca Raglon

Part 7. Self-Fashioning

12. The Spectacle of Science and Self: Mary Kingsley 215
Julie English Early

13. "Ape Ladies" and Cultural Politics: Dian Fossey and Biruté Galdikas 237
James Krasner

Part 8. The Tradition Continues

14. Interview with Diane Ackerman, 18 July 1994 255
Barbara T. Gates and Ann B. Shteir

Selected Bibliography 267
Index 271

Illustrations

3.1. Jane Haldimand Marcet 44
4.1. Agnes Mary Clerke 62
5.1. Catharine Parr Traill 80
6.1. Louisa Anne Meredith 99
6.2. *A Cool Debate* 109
7.1. Anna Botsford Comstock 117
7.2. "Hermit and Troubadour" 123
7.3. *Marumba modesta* 129
7.4. Luna moth 129
7.5. Plate VI from *Manual for the Study of Insects* 131
7.6. "Look at an insect . . ." 132
7.7. "Orchard Life" 133
7.8. John Henry Comstock 135
7.9. John Henry and Anna Botsford Comstock 139
8.1. Eliza Burt Gamble 148
9.1. An image of early mammals 167
9.2. Humankind and other hunters emerge after the Ice Age 168
9.3. First page of *The Fairy-Land of Science* 171
10.1. Charlotte Haldane 180

10.2. Naomi Mitchison 181

10.3. Here am "I" in the universe 189

10.4. Here am "I" in the past and future 190

10.5. Here are "we" in the world of values 191

11.1. Rachel Carson with binoculars 197

12.1. Mary Henrietta Kingsley 216

14.1. Diane Ackerman 256

Acknowledgments

The editors acknowledge contributions made to the development of this book by Marina Benjamin, Pamela Gossin, Robert Peck, and Alan Rauch. Ann B. Shteir's work on this book was funded by a grant from the Social Sciences and Humanities Research Council of Canada. Adrienne Auslander Munich extended much appreciated hospitality during the preparation of the manuscript. We appreciate the efforts of Suzanne Potts in preparing the manuscript for publication.

Stephen Jay Gould's essay is reprinted with the permission of *Natural History* magazine. Diane Ackerman's poem "The Dark Night of the Hummingbird" is reprinted with her kind permission. Portions of the introduction to this text, "Charting the Tradition," appeared in *Victorian Literature and Culture* 21 (1993) and portions of "Revisioning Darwin with Sympathy: Arabella Buckley" in *History of European Ideas* 19 (1994): 761–65.

Contributors

Diane Ackerman, poet, essayist, and naturalist, has published twelve books of poetry and nonfiction, including *A Natural History of the Senses* (1990) and, most recently, *The Rarest of the Rare: Vanishing Animals, Timeless Worlds* (1995).

Marianne Gosztonyi Ainley is Professor and Chair for the Women's Studies Programme at the University of Northern British Columbia. Her research areas include the history of Canadian women and scientific work, feminist scientific biography, First Nations women and environmental knowledge, and the history of ornithology. Her publications include *Despite the Odds: Essays on Canadian Women and Science* (1990).

Julie English Early, Associate Professor of English at the University of Alabama in Huntsville, has published work on late-nineteenth-century British and American ethnography, on scientific travel writing by Mary Kingsley, Margaret Fountaine, and Douglas Freshfield, and on the Anglo-Indian writer Flora Annie Steel. She has also written on Edwardian technology and crime, and is currently writing a book on Edwardian literature and culture.

Barbara T. Gates is Alumni Distinguished Professor at the University of Delaware. Among her publications are *Victorian Suicide: Mad Crimes and Sad Histories* (1988), *Critical Essays on Charlotte Brontë* (1990), and the *Journal of Emily Shore* (1991). She is currently at work on a study of Victorian women and nature.

Stephen Jay Gould is Professor of Geology and Alexander Agassiz Professor of Zoology at Harvard University. His many books include *Wonderful Life: The Burgess Shale and the Nature of History* (1989) and *Bully for Brontosaurus: Reflections in Natural History* (1991). He is a long-standing contributor to *Natural History* magazine.

Pamela M. Henson is Director of the Institutional History Division of the Office of Smithsonian Institution Archives. Her research focuses on the history of natural history, history of museums, and the role of women in science. Recent publications include "The Comstock Research School in Evolutionary Entomology," *Osiris* 8 (1993): 159–77; and with Terri A. Schorzman, "Videohistory: Focusing on the American Past," *Journal of American History* 78 (1991): 618–27.

Rosemary Jann is Professor and Co-Chair of English at George Mason University. She is the author of *The Art and Science of Victorian History* (1985) and *The Adventures of Sherlock Holmes: Detecting Social Order* (1995), as well as several recent articles on the construction of social difference by Victorian science.

Judith Johnston is an Arts Faculty Teaching and Research Fellow at the Department of English, University of Western Australia. She is author of the forthcoming *Anna Jameson: Victorian, Feminist, Woman of Letters* and is currently investigating Australian women's professional nonfiction writing, 1840–1910, with special focus on the publications of Louisa Anne Meredith.

James Krasner is Associate Professor of English at the University of New Hampshire. He is the author of *The Entangled Eye: Visual Perception and the Representation of Nature in Post-Darwinian Narrative* (1992) and is presently researching portrayals of women and animals in popular culture.

Bernard Lightman is Associate Professor of Humanities at York University. His publications include *The Origins of Agnosticism* (1987), *Victorian Faith in Crisis* (1990) (co-edited with Richard Helmstadter), and a number of essays on science, religion, and unbelief in Victorian England. Currently, he is writing a book-length study of popularizers of Victorian science.

Greg Myers is a lecturer in linguistics at Lancaster University, Lancaster, United Kingdom, where he is Director of Studies for the Programme in Culture and Communication. He has written *Writing Biology: Texts in the Social Construction of Scientific Knowledge* (1990) and *Words in Ads* (1994). He is now working on a study of talk about the environment.

Rebecca Raglon is an interdisciplinary scholar whose focus of research is the literary contributions women have made to environmental discourse. She teaches at the University of British Columbia. Her work has most recently appeared in *Alternatives, Environmental History Review,* and *Critique,* and a book of her short stories, *The Gridlock Mechanism* (1992), has been published by Oberon Press.

Ann B. Shteir is Associate Professor of Humanities and Director of the Graduate Programme in Women's Studies at York University. She is author of *Cultivating Women, Cultivating Science: Flora's Daughters and Botany in England, 1760 to 1860* (1996), and editor of Priscilla Wakefield's *Mental Improvement, or the Beauties and Wonders of Nature and Art* (1995). She edited "Women and Science," a special issue of *Women's Writing: The Elizabethan to Victorian Period* (1995). *Cultivating Women, Cultivating Science* was awarded the Joan Kelly Memorial Prize in Women's History for 1996 by the American Historical Association.

Susan Squier is Julia Gregg Brill Professor of English and Women's Studies at the Pennsylvania State University, University Park. She is the author of *Babies in Bottles: Twentieth-Century Visions of Reproductive Technology* (1994) and *Virginia Woolf and London: The Sexual Politics of the City* (1985), editor of *Women Writers and the City: Essays in Feminist Literary Criticism* (1984), and co-editor of *Arms and the Woman: War, Gender, and Literary Representation* (1989).

PART 1

CHARTING THE TRADITION

1

Introduction
Charting the Tradition

Barbara T. Gates and Ann B. Shteir

Prefatory

The essays in this book explore work by some of the many women who have disseminated scientific knowledge. They highlight women as productive literary and artistic agents within science culture and focus on science written in the vernacular. Since the late seventeenth century, women's books, essays, illustrations, and lectures have contributed significantly to the spread of scientific ideas. Women writers have worked in genres as varied as guidebooks, essays, juvenile fiction, and magazine journalism. Initially conceived as instruments of pedagogy, many of their works have been modeled on learners' needs. Using a range of narrative forms and rhetorical strategies, women have popularized science for general audiences and specialists alike. The art of women popular science writers as mediators of knowledge is a testimony to their capacity and tenacity; their story is a part of literary history, women's history, the history and sociology of science, and the history of education that needs to be re-explored and revalued. Relocating portions of that story is the primary task of this volume.

Recently, Roger Cooter and Stephen Pumfrey have detailed their reservations about the value of the umbrella term "popularization" in the history of science because, they suggest, it elides both science popularization and science in popular culture and is laden with negative and ideologically weighted connotations. They call for a model of popularization that will treat popular science as its own form of knowledge, shaped in relation to the needs of audiences beyond elite and learned culture. Sharing their interest in decentering negative approaches to science popularization, we hope that in accentuating and elaborating upon the encounter between the women who wrote science and their varied audiences, our volume will help to unpack some of the pejorative connotations surrounding the word "popularize."

In part because of an engrossing concern with the "woman question" in science and the "science question" in feminism, the extent to which scientific popularization added both to the diversity of scientific writing and to the diversity of women's writing has been largely overlooked. Instead, analysts of gender and science have focused more closely upon the social construction of terms like "science," "history," "nature," and "woman." These culturally determined categories, combined with changing ideas about what constitutes "masculine" and "feminine," have led at one time or another to the exclusion of women from scientific practice. Thus in recent years, feminist researchers on women, gender, and the history of women and science have amassed evidence to document ways in which such gendered thinking about women and science has limited access to scientific cultures.[1]

And while feminist researchers on gender and science have set about reinterpreting science culture, other feminist researchers within the fields of literary and cultural studies have been recuperating and analyzing women writers and their texts.[2] As a result, poetry, fiction, the art of letter writing, biography and autobiography, literary prose, historiography, and travel writing from the early modern period to the present day have in effect been recharted. Nevertheless, because women have been assumed to have inhabited only the distant borderlands of science, they have continued to be marginalized in most contemporary discussions of science writing. In attempting to rectify this marginalization, the essays in this book complement work in collections edited by Pnina Abir-Am and Dorinda Outram, Marianne Gosztonyi Ainley, and Marina Benjamin.

A Historical Contextualization

We call this section a historical contextualization rather than "the" historical contextualization because we are aware that we are introducing the history of only one culture: a culture of scientific popularization that anglophone British, North American, and Australian women built around the western science of the late-seventeenth to the late-twentieth centuries. This is scope enough; a broader history of popularization could include Europe, the scientific discoveries of the Arab world or the East, or the folk wisdom of peoples all over the globe. But the English-speaking world is one place to begin charting a tradition of women and popular science writing, for the Newtonian and Darwinian revolutions alone caused serious controversy and sufficient need for translation of scientific theories into the vernacular. The overview that follows does not attempt to cover the entirety of the three centuries. It begins with the Enlightenment and ends with the close of the nineteenth century. Nor is it comprehensive in terms of the diaspora of English speakers. There were, for example, popularizers in others of the Commonwealth nations and colonies, and trends and individuals that we have not been able to include here.

We begin with the Enlightenment because, although it has been critiqued in our day as the seedbed of European racism, imperialism, and scientism, it was also the historical period that brought women into western science culture. Science was placed on cultural agendas of improvement and self-improvement, and women participated in the production and dissemination of natural knowledge. At the end of the seventeenth century, Bernard de Fontenelle's witty and flirtatious conversations about astronomy in *Entretiens sur la pluralité des mondes* (1686) between a well-versed aristocratic philosopher and a marquise with a lively mind brought scientific learning to male and female polite audiences. Translated into English by the dramatist Aphra Behn as *Discovery of New Worlds* (1688), Fontenelle's book spawned a generation of popularizing works about science that found a wide readership (Douglas 1–14). In England, periodicals, scientific lectures, and books within polite culture promoted astronomy and other areas of natural philosophy. From the *Athenian Mercury* in the 1690s to the *Gentleman's Magazine* across the mid-century, print culture contributed to spreading interest in science. Audiences flocked to science demonstrations given by science lecturers who booked themselves into assembly rooms and staged public experiments about gravitation, magnetism, and electricity. Science increasingly became an activity of leisure for the landed gentry and the middle ranks of society. By the later decades of the eighteenth century, science culture also embraced young people and children as pupils, readers, and the audience for science as spectacle; and authors and publishers cultivated "the rising generation" as a market, with juvenile science titles such as *The Newtonian system of Philosophy, adapted to the capacities of young gentlemen and ladies* (Secord 127–51).

Mirroring the social and intellectual agendas of the Enlightenment, many books and periodicals in England introduced women to the popular sciences. The *Ladies' Diary,* a mathematical recreation magazine first issued in the early eighteenth century, was "designed on purpose for the diversion and use of the Fair Sex"; it included algebra, astronomy, fluxions, harmonics, and optics, and featured "enigmas and mathematical questions," puzzles and solutions contributed by readers (Perl 36–53). Benjamin Martin's *The General Magazine of Arts and Sciences* (1755–64) offered a series of articles entitled "The Young Gentleman and Lady's Philosophy," which were organized as conversations between a university student and his sister at home; these articles (later issued separately in several editions) were designed to familiarize readers with recent developments in astronomy, optics, and hydraulics.

Until the mid-eighteenth century women had participated in science mainly as audience. With the notable exception of Margaret Cavendish, whose theoretical works included *Observations upon Experimental Philosophy* (1666) and *Grounds of Natural Philosophy* (1668), they were most likely to have been readers and students at public lectures and demonstrations rather than shapers of knowledge. But the appearance of science topics during the 1740s in Eliza Haywood's magazine *The Female Spectator* signaled the beginning of women's using print culture to promote science for women. *The Female Spectator* (1744–45), one of the earliest periodicals written by women and for women, aimed to improve the morals and manners of its age through essays, stories, letters from putative correspondents, and editorial replies designed to enlarge the horizons of its female readers. The magazine's contributors recommended scientific activities relating to plants and insects, and presented natural history and natural philosophy as resources for women's amusement and improvement.

But *The Female Spectator* also promulgated gendered notions about women and knowledge. Eliza Haywood warned women against veering too much toward study and expertise, and wrote: "it is not my ambition to render my sex what is called deeply learned" (3:145). A contributor recommended that women undertake only "less severe and abstruse" work observing herbs, ants, and caterpillars, rather than "fill their heads with the propositions of an Aldrovanus, a Malbranche, or a Newton" (3:134, 125). Another periodical from Enlightenment England, *The Lady's Museum* (1760–61), likewise situated scientific content inside norms of gender ideology. Issued by the novelist Charlotte Lennox, it combined moral tales with translations from the French, instruction in science, and exhortations to women to improve their minds. A series of articles entitled "Philosophy for the Ladies" discussed topics in natural philosophy, and a series of essays about natural history featured ants, animal camouflage, and the metamorphosis of butterflies. Lennox did not want to shape a generation of learned ladies and women scientists by using her writing to teach natural philosophy in depth. Instead, her wish was "to render the ladies though learned not pedantic, *conversable rather than scientific*" (1:130).

During the second half of the eighteenth century, gender ideology, educational ideas, and publishing history converged, resulting in a cultural paradox. Writers across a wide political spectrum agreed that women's education was desirable, but agreed equally that female education should prepare women for domestic responsibilities. Some writers cautioned against women's learning or knowing too much—too much reading, too many languages, too much science. Phrasing their cautions in terms of dichotomies, they contrasted those women who sought too much knowledge or the wrong sort of knowledge with others who chose more appropriate domestic pursuits. Cultural discourses constructed women's bodies as maternal rather than sexual, and, Ruth Perry has argued, "colonized" women's bodies for domestic life and for heterosexual productive relations. Magazines, novels, and courtesy literature conflated maternal practice with femininity, and guided readers toward new familial and maternal missions. Science activities and science writing became part of the middle-class project to shape appropriate learning for women within a new-style family. As women across the middle ranks of society were directed toward home, family, education, and the general culture of piety and improvement, education of the young took on enlarged cultural value, and educational writings became an arena that women could claim for themselves. Women's science writing developed from this conjunction of complex social and cultural values.

Women writers of the later eighteenth century, having incorporated science into popular magazines (and also into courtesy books such as Ann Murry's *Mentoria* and *Sequel to Mentoria*), crossed the threshold into direct expositions of science by writing introductory books for children and parents. Their books about astronomy, chemistry, natural philosophy, entomology, and botany form an important part of the early textbook tradition in science. The best-known pioneer was Sarah Trimmer, a religious educator anxious to counteract "infidel books." Her *Easy Introduction to the Knowledge of Nature, and Reading the Holy Scriptures, adapted to the capacities of children* (1780) yoked natural knowledge and scriptural doctrine in a "general survey of the Works of Providence" (vii). Trimmer shaped a narrative in which a mother introduces two young people to information about plants, animals, and insects. As they walk together, she comments upon the natural objects around them—peacocks, butterflies, silkworms—and these topics in turn lead to larger moral and spiritual reflections. The book combines a tone of improvement with a style that is familiar and sweet; "Now you see, my Dears, that every thing, when we examine it, is curious and amusing;—None need go sauntering about, complaining that they have nothing to divert them, when they may find Entertainment in every Object in Nature" (40). Margaret Bryan likewise incorporated a familiar tone in *A Compendious System of Astronomy* (1797) and *Lectures on Natural Philosophy* (1806), textbooks based upon her experience teaching science to girls at schools in London and Margate. Her books project a caring and enthusiastic voice, contain first-person references, and incorporate moral and religious reflection. A frontispiece showing "Mrs. Bryan and her Children" represents the maternal manner of her science books. Like many other popular science books of her day, Bryan's astronomy book taught more than substantive scientific information. Her "grand design," she writes, was "to excite in my dear pupils a spirit of enquiry" (48), and many expository lectures end with a thought about virtue.

During 1780–1830, women science writers often chose letters, dialogues, and conversations—known as the "familiar format"—for teaching science to young readers and to women of all ages. Priscilla Wakefield used an epistolary form for *An Introduction to Botany* (1796), where letters between sisters taught rudiments of the Linnaean system of botany, and in *Mental Improvement, or the Beauties and Wonders of Nature and Art* (1794–97), she offered after-dinner family conversations in which a mother and father give their children informal instruction on a range of scientific and technological topics. Such conversational exchange exemplifies a family that cares for the intellectual and moral welfare of its children. Jane Marcet, who enjoyed particular prominence among writers of scientific dialogues (and who is discussed by Greg Myers later in this volume), also used a similar format in *Conversations on Chemistry* (1806) and *Conversations on Natural Philosophy* (1819). Popular science books such as Marcet's transformed French aristocratic popular science into popular and improving hobbies for the middle ranks of society in eighteenth-century England. They incorporated moral and spiritual teaching, featured family settings and narratives, and digressed into topics of more general interest. Shaped by women writers, they merged gender with genre, and became pioneering contributions by women to the cultural transmission of science.

Science learning was thus not only associated with schoolwork but also represented as part of daily life and the general education of young people. Epistolary and conversational narratives about science illustrated a pedagogy of interpersonal connection, a teaching climate that modeled a new style of family practice. This contrasts with another popular textbook style at that time, the catechistic style of question and answer, which teaches science without personalities and without context, and presupposes little contact between teacher and pupil. Richmal Mangnall's *Historical and Miscellaneous Questions for the Use of Young People* (1800), for example, had been written for pupils "in public Seminaries" whose teachers do not have "sufficient leisure to converse with each separately" (ii). The learning pattern of the catechism represents instruction rather than education, and school culture rather than improving leisure activities.

Most women science writers of the late eighteenth century used the narrative form of conversations, dialogues, or letters in their books, and featured families, domestic settings, and characters with names and personalities. Mitzi Myers has described a "female mentorial tradition" in children's literature in eighteenth-century England, in which a mother, aunt, or governess is invested with power and authority to initiate her charges into knowledge of the world. This important element in children's literature appears in women's science writing as well. In many popular science books by women written between the 1780s and the 1840s, mother (or a mother-substitute) taught science to her children. Her interest in science was depicted as being part of her other responsibilities, and science education was presented as a part of good mothering. The maternal science teacher, then, served as a figure of power and expertise for her children, and the Scientific Mother was also an exemplar of female knowledge and intellectual authority for adult readers.

Nevertheless, as scientific discoveries continued to unfold in the nineteenth century, science became increasingly masculinized and professionalized. In their books on the gendered nature of science, both Carolyn Merchant and Ludmilla Jordanova have forefronted a telling 1899 representation of "Nature unveiling herself before Science," a statue executed by Louis Ernest Barrias that shows a young female figure unveiling herself (Merchant 190–91; Jordanova, *Sexual Visions* chap. 5). Elaine Showalter has in turn suggested that if this image could only have been reversed in a companion piece entitled *Science Looking at Nature*, "it would have depicted a fully clothed man, whose gaze was bold, direct, and keen, the penetrating gaze of intellectual and sexual mastery" (*Sexual Anarchy* 145). By the mid-nineteenth century, the Scientific Mother was edged out by men of science who seem to have appropriated the gaze as well as the words of women. Nevertheless, women did look and they did speak, creating a counterdiscourse not just about nature but also about science.

Because the British cultural and educational bias so strongly favored men in the nineteenth century, none of the major discoveries in natural and physical science belonged to women. Yet many women took an intense interest in those discoveries and throughout the century sought knowledge of the workings of the universe, often at the urging of other women. Botany, the fern craze; geology, the rock-collecting craze; entomology, the bug-hunting craze—all of these became female more than male pursuits. Charles Kingsley's *Glaucus* (1855), a book written to inform a hypothetical London merchant about the wonders of the seashore, asks the merchant to look at what pleasures his daughters have gained from their "pteridomania" over ferns and to imagine what equivalent joys he himself might find in a study of the seashore (4–5).

In the nineteenth century educated women interested in science still formed large portions of the audience at public lectures and read whatever was available to them by way of written explanation. Even more than in the previous century, they also retold the story of science. In the system of scientific practices defined by their culture, they functioned not as the groundbreakers but as educators, carefully explaining new views of the physical and natural world to women, children, and the working classes. Typically, the scientific theories they conveyed were those that had become accepted; they eschewed the controversial partly in order to enhance their authority. Their originality lay not in the substance of what these women were trying to convey but in the distinctive discourse that they evolved as they narrated the story of science.

We do not wish to suggest that there was a gender war between scientific texts of the nineteenth century—to search out and stress binary oppositions between male and female scientific narratives. We do not think that for the most part the women popularizers believed they were hampered by their gender: their writing gave them voices and in many cases livelihoods as well. These women found their audience, and they seemed to like the job of informing it. And they were appreciated: up until the end of the nineteenth century, their volumes sold as well as men's. When George Eliot and George Henry Lewes arrived in Ilfracombe in 1856, awkward and ill-equipped novices in seashore life but ready to learn enough for Lewes to be able to write *Seaside Studies* in 1858, they would have been as likely to have been carrying Anne Pratt's *Things of the Sea Coast* (1850) as they would Philip Gosse's *Aquarium*

(1854) or Kingsley's *Glaucus*. Both men and women popularized science, but women established a set of narrative paradigms that they made their own and that are revealing of women's writing and women's authority in nineteenth-century culture.

Many turn-of-the-century writers of dialogue were, for example, motivated by a commitment to natural theology. In describing the wonders of nature, they believed they were also describing the wonders of creation and, ultimately, of a creator. Most were equally motivated by a belief in the accessibility of proof. For Priscilla Wakefield in *Domestic Recreation; or, Dialogues Illustrative of Natural and Scientific Subjects* (1805), "the curious phenomena that nature presents, is [*sic*] one of the most rational entertainments we can enjoy: it is easy to be procured; always at hand; and, to a certain degree, lies within the reach of every creature who has the perfect use of his senses, and is capable of attention" (77–78). For women like Wakefield, moral education and scientific observation were not at odds. On the contrary, observing and teaching about the natural world amounted to a calling. Since women saw themselves as the moral educators of each other and of the young, it only made sense that they should be the ones to offer a proof-laden version of natural theology to women and children. They became important purveyors of the narrative of natural theology.

In *Writing Biology*, Greg Myers establishes two categories in order to describe twentieth-century science writing: the "narrative of natural history," which refers to a popular account of nature that is diverting, full of anecdotes, and nontheoretical—the sort of thing one finds in natural history magazines—and the "narrative of science," which describes a work that must meet the standards of a discipline and is heavily committed to model building (142–43, 194–96). The purpose of the latter is primarily to establish the credibility of a scientist within the scientific community. Students of the nineteenth century will notice that Myers's distinctions can easily apply to the Victorian era; Thomas Henry Huxley and many mid- to late-century women popularizes, for example, utilized the "narrative of natural history"; Charles Darwin the "narrative of science." But at the turn of the century, William Paley, not Darwin, was the naturalist to be reckoned with. His *Natural Theology* (1802) claimed that each discovery of natural science was new proof of the wisdom and power of a divine creator. For half a century, women took the substance of Paley's work as the stuff of their popularizations of natural history, producing narratives of natural theology. But by the 1840s, both the narrative of natural theology—an appeal to the importance of William Paley in the work of early women popularizers of science—and the widespread use of dialogue as the appropriate vehicle for women's popularizations of science were diminishing. Evolution was in the air; Paley was being displaced by new scientists like Sir Charles Lyell, Alfred Russel Wallace, and Charles Darwin; and the dialogue, which came of age along with natural theology, was outliving its own credibility. People were, of course, better informed scientifically, partly because of the dialogues themselves. But learning by question and answer was more frequently satirized—by Charles Dickens, for example. In *Our Mutual Friend,* his Miss Peecher and Mary Anne fall into a mock dialogue of "learning." Wherever she goes, whatever she is asked, young Mary Anne reels off the expected rote answer, so inured is she to interrogation.

For half a century the scientific dialogue had suited women practitioners. Certainly it established women as scientific authorities. Questions imply answers, and authorities of one sort or another are needed to supply them. When in 1869 Charles Kingsley decided to write a popular book of lessons in geology for children, he titled it *Madame How and Lady Why.* Kingsley's book, addressed to young boys, made these two female figures aspects of Nature herself, the ultimate authority on how and why. The real-life female instructors were experts, too. Like all good educators, they seem to have studied their subject well enough to translate it from professional jargon into the vernacular and to cull just the right examples from their store of knowledge to impress their pupils with self as well as science.

By mid-century, new forms began to enhance, and ultimately would replace, the conversation and dialogue. In her *Young Naturalist's Journey* (1840), for example, Jane Loudon offered a type of narrative that would widen the scope of popular science writing. In Loudon's "journey," a fictional young girl and her mother travel throughout the British Isles and meet, query, and parley with people who own or have extensive knowledge of exotic animals. Loudon's is a narrative of natural history, not natural theology, and is informed by other vehicles for popularizing science. It was when she was turning over the pages of the *Magazine of Natural History* looking for information about animals that Loudon was stimulated to write her own book. "Stripped of their technicalities," she believed the magazine articles could "be rendered both interesting and amusing to children" (ix). She went on to carefully authenticate her own originality, adding that "the adaptation of them has cost me quite as much time and labour as the writing of an entirely new work. . . . anecdotes here related of the animals are strictly true; though the incidents of the journey, and the persons introduced are partly imaginary" (x).

Loudon patterned her main characters after her own daughter and herself, producing this book and dozens of others in order to support the two of them after the death of her husband, John Claudius Loudon the landscape gardener. Her journey is a railroad journey, not a turn-of-the-century ramble in nearby rural surroundings, and with Loudon we enter wider worlds in terms of both natural history and geography. In this regard, Loudon was best known for her botanicals, like *The Ladies' Flower Garden of Ornamental Annuals* (1840) and *British Wild Flowers* (1846), which were part of a movement toward botany that had attracted British women as illustrators and enthusiasts throughout the second quarter of the nineteenth century. These books were designed to familiarize women with the plants of the British Isles, plants they could see with their own eyes and grow with their own hands. But a still wider world of botany was opening to reading and traveling women, and they in turn tried to interest other women in that world. Elizabeth Twining's beautifully worked *Illustrations of the Natural Orders of Plants with Groups and Descriptions* (1849–55) was prefaced to promote this new catholicity of vision. Twining, whose family owned tea plantations in many corners of the English empire, wanted to show British plants in relation to others. "By thus placing our native plants in groups with foreigners, we acquire a more correct idea of the nature of our Flora, and the character it has when compared with that of other countries" (ii), she explained. Proud of her pioneering efforts at popularizing comparative botany, she helped her readers to see their favorites in a new light. Loudon and earlier popularizers had sympathized with women's difficulties with the Linnaean system of classification. Twining chose an even more difficult system to follow—Candolle's—because his ordering allowed her readers "a ready perception of the geographical distribution of any particular tribe. Also what proportion our British Flora bears, both in quantity and quality, to the whole range of the Natural Orders" (ii).

By mid-century other women popularizers had already seized upon this outward expansion in search of wider worlds, possibly because, as women, they felt the constrictions of both British society and the British scientific community. Becoming pioneers and going beyond British borders offered women horizons against which to shine as authorities and granted them a stature more equivalent to that of men. Rosina M. Zornlin, author of over half a dozen books popularizing science in the 1850s—among them *Outlines of Geology* (1852), *Recreations in Geology* (1852), and *The World of Waters* (1855)—certainly opted to commend the new vantage points. In her *Recreations in Physical Geography, or the Earth As It Is* (1855), she praised the work of travel writers and showed how it applied to her own work: "descriptions of small and detached portions of the earth's surface kindle in us a desire to become acquainted with all that is remarkable on the face of the globe: in short, with all that descriptive geography . . . can impart to us" (4). This desire then led to a need to understand physical geography, and hence to her book.

Audiences for scientific writing had become more demanding and more sophisticated in high Victorian Britain. Writing on behalf of the reader or student of science, in *The Scientific Lady* Patricia Phillips explains that "women were no longer content to be supplied with their science in a haphazard manner" (191). Nevertheless, female devotion to the moralizing of scientific discovery also remained. Zornlin, so aware of the far-ranging, so devoted to getting an accurate picture of geology and geography to the working classes—"to the cotter no less than the large landed proprietor, to the artisan as well as the master manufacturer, to the miner as well as the owner of mines" (preface to *Recreations in Physical Geography* vii)—and so convinced of the "truth" of science, nevertheless clearly stated that truth was "God's" truth (*Recreations in Geology* 365). She was aware of the latest discoveries in her fields, but opted to present them conservatively.

So did Margaret Gatty, writer of the authoritative *British Seaweeds,* completed in 1862 after fourteen years of painstaking research, who continued for decades to reissue editions of her *Parables of Nature* (1855), one of the most popular books of the last half of the nineteenth century. Unlike the *Seaweeds,* the parables, although based in scientific knowledge, were meant to illustrate morals of one sort or another and reverted to ideas of natural theology. In the parable "Knowledge not the Limit of Belief," for example, a zoophyte, a seaweed, a bookworm, and a naturalist parley about a controversy that had earlier interested Gatty: whether a zoophyte was or was not a plant, and whether a seaweed was an animal. The naturalist solves the problem, showing that despite popular opinion and the bookworm's protests, the zoophyte is animal, the seaweed a plant. Even the two creatures themselves are humbled by this discovery and become "disciples" of a higher power—the naturalist—as children must in turn become disciples of God.

Thus along with their challenge to popularize science in more sophisticated and less amateurish ways, women like Gatty and Zornlin felt they still bore a responsibility as religious or moral teachers. Other post-Darwinian women like Alice Bodington saw popularization in different terms. In her *Studies in Evolution and Biology* (1890), Bodington attempted to bring the story of evolution scientifically up-to-date for turn-of-the-century readers. Her unapologetic tone was a considerable departure from those of earlier women popularizers. "A stigma is supposed to rest," she complained, "for some mysterious reason, upon the person who ventures to write upon any branch of science without being an original discoverer. I am at a loss to imagine why it is considered almost wrong to write about physical science without having made original experiments. A historian is not required to have fought in the battles he describes, nor a geographer to have personally traversed the wilds of Africa. Why cannot a wide view be taken by some competent person of the results of the labours of hundreds of scientists, so that we may more clearly see what manner of fabric is being reared?" (preface, ix–x). This is a direct plea on behalf of the expansiveness of the narrative of natural history rather than the narrative of science—the more narrow, male preserve. Bodington wanted to be able freely to enter into the shared preserve without the condemnation of scientists. She especially wanted to be unobstructed in her attempt to supplement Darwin's narrative where Darwin seemed passé.

She made this clearest in her essay "The Mammalia: Extinct Species and Surviving Forms," where she took some exception to Darwin's idea of natural selection. The "great master," as she calls him, "had only grasped one form of the law governing evolution . . . whereas we now see that the infinite, delicate variations in the world of organic beings are owing to the intense irritability and susceptibility to molecular changes of protoplasm, and the consequent action of the environment upon it" (22). What she objected to in Darwin was his determinism. To her, "natural selection evoked some unknown force vaguely of the nature of will. The action of the environment upon protoplasm requires nothing but ordinary and well-known phenomena of organic chemistry" (23). Bodington preferred this explanation because it gave natural selection a lesser role in "the great drama of development" (23).

Bodington not only identified but pinpointed her intended audience—nonprofessionals who do not realize that, despite the many discoveries since Darwin's time, obscurities still veil the story of "the appearance, duration, and disappearance of species" (60). This permitted her to urge her readers, especially the travelers and anthropologists among them, to continue hunting for missing links. In its choice of audience, Bodington's post-Darwinian work makes an interesting contrast with Arabella Buckley's and Eliza Burt Gamble's, both of whom are discussed in essays in this volume. Bodington was clearly not writing for children, nor specifically for male or female readers, but for what she considered a general audience of adults. She wanted to make "clear and plain" to persons of "ordinary intelligence" those many new observations of scientists who "buried" their work "in the pages of scientific journals, to be read only by specialists" (143). All the same, she could see no reason not to make the narrative of natural history as amusing and pleasurable as possible.

Nor, at the end of the century, could many others. Bodington worked the same vein as did male popularizers of science like J. G. Wood. Although women like Arabella Buckley and Mary Kingsley lectured on the subjects in which they had expertise, science lecturing was predominantly the domain of men, whose lectures helped them to build reputations and to sell books. Wood himself developed a unique art that he called "sketch-lecturing." He would speak, then illustrate his subject before his audience's eyes, dramatically conjuring up huge images of hydrozoa, fish, or other animals in colored pastilles, which he applied directly to a large black canvas. When he died, the year before Bodington's *Studies* appeared, Wood was heralded as the great popularizer of his day; when he wrote his *Boy's Own Book of Natural History* in 1883, the year of Arabella Buckley's *Winners,* he himself claimed that there was "no work of a really popular character in which accuracy of information and systematic arrangement are united with brevity and simplicity of treatment" (iii).

Yet women like Buckley and Bodington did rival him, and they were both supporters of the theory of evolution that he, with his conservative religious bias, had trouble condoning. By the end of the century, women, long purveyors and champions of the popularization of science, were now more well-informed, original, and creative popularizers than they had been in the past. But they were also perceived as competitors in a literary struggle for survival and were more frequently discounted by men like Wood. Edmund Gosse, for example, in a supposed tribute to the popular natural history writer Eliza Brightwen that introduces her posthumous *Life and Thoughts of a Naturalist* (1909), all but kills his subject with condescension. "This little book," he says of Brightwen's autobiography as he compares it with her natural history, "in its simplicity, in its *naiveté,* will not be comprehended by any but those who are already in sympathy with its author and in measure conversant with her methods" (x). By implication, then, those methods were unsophisticated, rather like the young women's "pteridomania" described by Kingsley in *Glaucus.* Kingsley had minimized the young women's pursuit in order to make his intended audience, the London merchant, feel superior to his daughters. They had taken up ferns only in preference to "the abomination of 'Fancy work,' that standing cloak for dreamy idleness" (4), whereas the merchant should take up the study of seashores as an amateur naturalist. As the century wore on, botany, along with the seashores, was becoming the province of men, although it had always been the science most attractive to women. In

1887, J. F. A. Adams's essay in *Science,* "Is Botany a Suitable Study for Young Men?" offered botany as an area to be studied by "able-bodied and vigourous brained young men" (116) as well as by women. Adams's attitude, like Kingsley's and Gosse's, suggests that all scientific endeavor had become linked to masculine intellectual qualities and that women had begun to lose their status as interpreters of the narrative of natural history.

In part this loss of status also occurred because their special audience of children and women had begun to diminish. Natural history education had reached the schools—first girls' schools, then boys'—and had reduced the need for science books in the home. In the first half of the century, mothers and grandmothers had bought books by women popularizers and had virtually educated themselves along with their children. But as time went on, mothers and grandmothers were progressively less responsible for children's knowledge of science. Concurrently, formal education in natural history for girls, an important ingredient in their curricula for decades, was lessened, while at the same time it was increased for boys. Young men became more knowledgeable in science, more likely to learn their science from men, and more and more likely to be the future narrators of the narrative of natural history as well as the narrative of science. This overall "demise" of science as a female interest, as Patricia Phillips describes it in chapter 8 of *The Scientific Lady,* was especially noticeable after the Taunton Commission's report in 1868–69. It was then that the focus of women's education shifted from natural science to the classics, ostensibly to make women's study equivalent to men's. By the time of the Bryce Commission report in 1894–95, many of the bright young women in secondary schools who showed a special desire to study science were discouraged from taking classes in scientific subjects and encouraged instead to work to perfect a knowledge of classics. Thus in the name of equalizing formal education for women, fewer young women were given access to narratives of science in secondary schools, while at almost the same time, fewer were reading narratives of natural history in their homes. Ironically, then, the economics of the book trade and the masculinization of science—through appropriation by and education of men—joined with educational reforms on behalf of women to prevent women from freely retelling the story of science.

To Introduce the Essays

The essays in this volume are intended to give the reader a sense of various aspects of women and science in the vernacular, and are arranged accordingly. Important figures like Mary Somerville and Harriet Martineau, fiction writers as different as Mary Shelley and Beatrix Potter, and dozens of other women on every side of every ocean are noticeably missing. We have left them to future investigators. Although questions of agency, dissemination, and production are critical to the essays, questions of audience enter only tangentially. In our estimation, the complex subject of the audiences for works of popular science demands further research and another volume. What we offer here is a grouping of women, not a phalanx. Some of the women follow conservative lines and promote ideas already offered by others. Other women, like Rachel Carson, dare to break new ground. All of them repackaged science in one way or another. Equally, our contributors have packaged their own essays in diverse ways, looking to archival materials about writers, or focusing upon a telling moment or text that connects to broader issues in science culture.

Work in the area of women and scientific popularization often begins with efforts of recuperation. As historian and popularizer of science Stephen Jay Gould well realizes, the profiles of many little-known practitioners of our art have remained invisible. Gould steps into the field and opens our series of essays by recording his own encounter with books by Mary Roberts, a versatile mid-nineteenth-century writer whose work ranged over many areas of natural history. The conservative scientific content of Roberts's *Conchologist's Companion* first caught Gould's attention. He began by reading Roberts through a gendered scrim—as timid, bounded—but went on to glimpse rebellion and anger in her writing that led him to ask a larger question about women popular science writers of Roberts's time: "What really went on behind the mask of acceptance and convention respected by most women writers on natural history?" As Gould's essay suggests, recuperating early women science writers and revisiting their works can call for reading beyond historically constructed gendered assumptions about popular science books and their narratives.

Essays in the next section, "Disseminating Knowledge," offer discussions of ways in which women popularizers spread scientific knowledge to different kinds of audiences in different countries. Working with the rhetoric of science, Greg Myers's essay, "Fictionality, Demonstration, and a Forum for Popular Science: Jane Marcet's *Conversations on Chemistry*," sets Marcet's *Conversations* in its contemporary context. Myers compares Marcet to Humphry Davy and to a primary reviewer of her time, Henry Brougham, and highlights the new forum for science created by Marcet's conversations. Marcet's fictional dialogues, created to prepare women to understand the complexities of chemistry, domesticate and then dedomesticate science; they enabled women to grasp the essentials of chemistry in the home and prepared them to comprehend the rigors of the chemistry lecture at the Royal Institution.

From the time of Margaret Cavendish in the later seventeenth century, women were actively involved in reinterpreting the discoveries of astronomy. In "Constructing Victorian Heavens: Agnes Clerke and the 'New Astronomy,' " Bernard Lightman introduces a versatile and prolific late Victorian popularizer of science. Clerke, a new-style popularizer, compiled specialized scientific information not just for general readers but also for scientific experts. Her facility in crossing borders between the worlds of the general public and the professionals made her hard to categorize in her day. Was she a traditional popularizer? Was she a scientist?

Off in the new worlds of Canada and Australia, the British tradition of natural history writing continued to serve women intellectually, socially, and economically. As Marianne Gosztonyi Ainley shows in "Science in Canada's Backwoods: Catharine Parr Traill," Traill wrote with an eye to book buyers in Canada, America, and England, and produced books and magazine articles about flora and fauna for general readers. Her pioneering botany manuals and field guides displayed an expertise based on field experience. Traill wrote in the same mid-nineteenth-century timeframe as did Louisa Anne Meredith, whose work Judith Johnston discusses in "The 'Very Poetry of Frogs.' " As Johnston charts the evolution of Meredith's career both temporally and geographically, she uncovers the story of a one-time Englishwoman turned Australian who interprets colonial natural history both as a Britisher (an exotic) and as an insider (an Australian). Meredith's later work reveals more of the biases of the Englishwoman abroad when Meredith begins to interpret native peoples through the same appropriating eye she earlier cast upon plants and animals.

Half a century later in the United States, Anna Botsford Comstock had an even more varied career as a science popularizer than did Traill or Meredith. Comstock worked as an illustrator and a writer of books for both adults and children, and pioneered in providing study guides for the emergent nature study movement she fostered. She also wrote for popular magazines, and shaped tales about animal, insect, and plant life, often including poetry in her popular writing. Basing her pedagogical commitments in close observation and in love of nature, she believed in teaching science through visual means, and contributed drawings and wood engravings to scientific textbooks co-authored with her entomologist husband, Henry Comstock. Henson's work shows why, at this time in intellectual culture, it is more important to locate women science writers in terms of their rhetoric and chosen genres than in terms of their more traditional roles as helpmeets and research assistants.

As women historically helped disseminate scientific knowledge, so they also helped define and redefine it, sometimes by entering or even creating scientific controversy. Rosemary Jann's "Revising the Descent of Woman: Eliza Burt Gamble" highlights the Darwinian counternarrative of an American suffragette. Gamble's study in cultural evolution, *The Sexes in Science and History,* is a maternalist account that features altruism rather than control as the motive force of evolution. Altruism is also a focus of Barbara Gates's essay "Revisioning Darwin with Sympathy: Arabella Buckley." Gates examines this British author's admiration and interpretation of Darwin but stresses Buckley's critique of Darwin on mutual aid. Buckley, whose books for children anticipated Kropotkin by twenty years, chose to stress species cooperation along with the struggle for survival.

Post-Darwinian discussions of women and their biological importance continued on after Darwin's century, particularly in the wake of the eugenics movement and debates over birth control. Susan Squier makes this abundantly clear in her essay, "Conflicting Scientific Feminisms: Charlotte Haldane and Naomi Mitchison," which takes as the site of its argument the wife and sister of J. B. S. Haldane. Squier contrasts the two as representatives of divergent feminist views of women as reproductive/sexual bodies. Whereas Charlotte Haldane, like many of her predecessors among women popularizers, valorized the maternal role, Naomi Mitchison worked to illustrate an expanded range of biological and social possibilities for women. Haldane became a scientific meliorist; Mitchison challenged the disciplinary and epistemological premises of scientific practice.

Radically different from both was the American Rachel Carson, best known for her Cassandra role as author of *Silent Spring.* Rebecca Raglon's work in "Rachel Carson and Her Legacy" resets the poetics of Carson's science writing in the context of nature writing. Carson deliberately wrote in the vernacular, intentionally diverging from more technical and scientific language, styles, and vehicles in much of her work. Risking reputation sometimes to the point of attracting derision, Carson reshaped environmental thinking in her popular scientific works; she redefined knowledge as she resculpted nature writing. Raglon's essay positions Carson as a foremother of contemporary environmental writing by women in Europe, the United Kingdom, and the United States.

While Carson chose to diverge from accepted models for scientific presentation in order to reach a wider audience, other women have felt an urgent need to authorize their voices as experts in order to be heard at all. Victorian Mary Kingsley, for example, refused the invisibility of Mary Roberts, the modest voice that enshrines the gender ideology of "true womanhood." As Julie English Early argues in "The Spectacle of Science and Self: Mary Kingsley," Kingsley constructed a stage persona that bears comparison with the distinctive voice of her travel writing. Kingsley's self-promotion, which packed Victorian lecture halls, served to enable her to impart the scientific information gleaned when she hunted fish and fetish in West Africa. By the later twentieth century, such self-fashioning could become treacherous. In " 'Ape Ladies' and Cultural Politics: Dian Fossey and Biruté Galdikas," James Krasner connects writings by the two primatologists to the changing political climates of the past two decades. Examining representations of Fossey and Galdikas in visual accounts in *National Geographic,* he reveals how such accounts utilize popular narrative plots about women, nature, and scientists.

Our book concludes with the voice of Diane Ackerman, poet, naturalist, and essayist. In conversation with the editors of this book, Ackerman muses and discourses upon popularizing science, now and in the past. Dedicated to the importance of what she calls "communion with nature," Ackerman approaches and represents the universe by blending art and science, poetry and prose. Our conversation reveals ways in which the strong tie between women and scientific popularization persists in our own day. We believe that in its content, purpose, and direction it echoes the familiar conversational scientific narratives of the eighteenth and nineteenth centuries and embodies the tradition that the essays in this collection begin to chart.

Notes

1. On gender and scientific practice, see Keller, *Reflections* and *Secrets of Life;* Harding, *Science Question;* Haraway, "Game of Cat's Cradle" and *Primate Visions;* Jordanova, "Gender and the historiography of science" and *Sexual Visions;* Merchant; Schiebinger, *Nature's Body* and *The Mind Has No Sex?*

2. To mention just a few among many notable examples of recuperative work in literary studies, see Showalter, *Literature of Their Own;* Todd; Ezell; Blain, Grundy, and Clements; Hobby.

Works Consulted

Abir-Am, Pnina G., and Dorinda Outram, eds. *Uneasy Careers and Intimate Lives: Women in Science, 1789–1979.* New Brunswick: Rutgers UP, 1987.

Adams, J. F. A. "Is Botany a Suitable Study for Young Men?" *Science* 9 (1887): 116–17.

Ainley, M. G., ed. *Despite the Odds: Essays on Canadian Women and Science*. Montreal: Véhicule P, 1990.
Benjamin, Marina, ed. *A Question of Identity: Women, Science, and Literature.* New Brunswick: Rutgers UP, 1993.
Benjamin, Marina, ed. *Science and Sensibility: Gender and Scientific Enquiry, 1780–1945.* Oxford: Basil Blackwell, 1991.
Blain, Virginia, Isobel Grundy, and Patricia Clements, eds. *The Feminist Companion to Literature in English: Women Writers from the Middle Ages to the Present.* New Haven: Yale UP, 1990.
Bodington, Alice. *Studies in Evolution and Biology.* London: Elliot Stock, 1890.
Brightwen, Eliza. *Life and Thoughts of a Naturalist.* London: T. Fisher Unwin, 1909.
Bryan, Margaret. *A Compendious System of Astronomy: A Course of Familiar Lectures; in Which the Principles of That Science Are Clearly Elucidated, So As to Be Intelligible to Those Who Have Not Studied Mathematics.* London, 1797.
Bryan, Margaret. *Lectures on Natural Philosophy, the Result of Many Years Practical Experience of the Facts Elucidated.* London, 1806.
Buckley, Arabella. *The Fairy-Land of Science.* London: Edward Stanford, 1879.
Buckley, Arabella. *Moral Teachings of Science.* London: Edward Stanford, 1891.
Buckley, Arabella. *A Short History of Natural Science.* London: John Murray, 1876.
Buckley, Arabella. *Winners in Life's Race.* London: Edward Stanford, 1883.
Cooter, Roger, and Stephen Pumfrey. "Separate Spheres and Public Places: Reflections on the History of Science Popularization and Science in Popular Culture." *History of Science* 32 (1994): 237–67.
Davidoff, Leonore, and Catherine Hall. *Family Fortunes: Men and Women of the English Middle Class, 1780–1850.* Chicago: U of Chicago P, 1987.
Douglas, Aileen. "Popular Science and the Representation of Women: Fontenelle and After." *Eighteenth-Century Life* 18 (1994): 1–14.
Ezell, Margaret J. M. *Writing Women's Literary History.* Baltimore: Johns Hopkins UP, 1992.
Fontenelle, Bernard de. *Conversations on the Plurality of Worlds.* Trans. H. A. Hargreaves, with introduction by Nina Ratner Gelbart. Berkeley: U of California P, 1990.
Gates, Barbara T. "Retelling the Story of Science." *Victorian Literature and Culture* 21 (1993): 289–306.
Haraway, Donna J. "A Game of Cat's Cradle: Science Studies, Feminist Theory, Cultural Studies." *Configurations: A Journal of Literature, Science, and Technology* 1 (1994): 59–71.
Haraway, Donna J. *Primate Visions: Gender, Race, and Nature in the World of Modern Science.* New York: Routledge, 1989.
Harding, Sandra. *The Science Question in Feminism.* Ithaca: Cornell UP, 1986.
Harding, Sandra, and Jean F. O'Barr, eds. *Sex and Scientific Inquiry.* Chicago: U of Chicago P, 1987.

Haywood, Eliza. *The Female Spectator* (1744–45). Glasgow, 1775.
Hobby, Elaine. *Virtue of Necessity: English Women's Writing, 1649–88*. Ann Arbor: U of Michigan P, 1989.
Jordanova, Ludmilla. "Gender and the Historiography of Science." *British Journal for the History of Science* 26 (1993): 469–83.
Jordanova, L. J. *Languages of Nature: Critical Essays on Science and Literature*. New Brunswick: Rutgers UP, 1986.
Jordanova, Ludmilla. *Sexual Visions: Images of Gender in Science and Medicine between the Eighteenth and Twentieth Centuries*. Madison: U of Wisconsin P, 1989.
Keller, Evelyn Fox. *Reflections on Gender and Science*. New Haven: Yale UP, 1985.
Keller, Evelyn Fox. *Secrets of Life, Secrets of Death: Essays on Language, Gender, and Science*. New York: Routledge, 1992.
Kingsley, Charles. *Glaucus; or, The Wonders of the Shore*. Cambridge: Macmillan, 1855.
Lennox, Charlotte. *The Lady's Museum*. London, 1760–61.
Loudon, Jane. *The Young Naturalist's Journey; or, Travels of Agnes Merton and Her Mother.* London: William Smith, 1840.
Mangnall, Richmal. *Historical and Miscellaneous Questions for the Use of Young People*. London, 1800.
Marcet, Jane. *Conversations on Chemistry*. London, 1806.
Marcet, Jane. *Conversations on Natural Philosophy*. London, 1819.
Martin, Benjamin. *The General Magazine of Arts and Sciences*. London, 1755–64.
Merchant, Carolyn. *The Death of Nature: Women, Ecology, and the Scientific Revolution*. San Francisco: Harper & Row, 1980.
Meyer, Gerald Dennis. *The Scientific Lady in England, 1650–1760: An Account of Her Rise, with Emphasis on the Major Roles of the Telescope and Microscope*. Berkeley: U of California P, 1955.
Murry, Ann. *Mentoria: or, The Young Ladies Instructor.* London, 1778.
Murry, Ann. *Sequel to Mentoria*. London, 1799.
Myers, Greg. "Science for Women and Children: The Dialogue of Popular Science in the Nineteenth Century." In Shuttleworth and Christie.
Myers, Greg. *Writing Biology: Texts in the Social Construction of Scientific Knowledge*. Madison: U of Wisconsin P, 1990.
Myers, Mitzi. "Impeccable Governesses, Rational Dames, and Moral Mothers: Mary Wollstonecraft and the Female Tradition in Georgian Children's Books." *Children's Literature* 14 (1986): 31–59.
Paley, William. *Natural Theology; or, Evidences of the Existence and Attributes of the Deity, Collected from the Appearances of Nature*. London: Faulder, 1802.
Perl, Teri. "The Ladies' Diary or Woman's Almanack, 1704–1841." *Historia Mathematica* 6 (1979): 36–53.
Perry, Ruth. "Colonizing the Breast: Sexuality and Maternity in Eighteenth-Century England." *Journal of the History of Sexuality* 2 (1991): 204–34.
Phillips, Patricia. *The Scientific Lady: A Social History of Woman's Scientific Interests, 1520–1918*. London: Weidenfeld and Nicolson, 1990.
Plues, Margaret. *Rambles in Search of Ferns*. London: Houlston, 1861.

Poovey, Mary. *The Proper Lady and the Woman Writer.* Chicago: U of Chicago P, 1984.
Schiebinger, Londa. *The Mind Has No Sex? Women in the Origins of Modern Science.* Cambridge: Harvard UP, 1989.
Schiebinger, Londa. *Nature's Body: Gender and the Making of Modern Science.* Boston: Beacon P, 1993.
Secord, James. "Newton in the Nursery: Tom Telescope and the Philosophy of Tops and Balls, 1761–1838." *History of Science* 23 (1985): 127–51.
Shevelow, Kathryn. *Women and Print Culture: The Construction of Femininity in the Early Periodical.* London: Routledge, 1989.
Showalter, Elaine. *A Literature of Their Own: British Women Novelists from Brontë to Lessing.* Princeton: Princeton UP, 1977.
Showalter, Elaine. *Sexual Anarchy: Gender and Culture at the Fin-de-Siècle.* New York: Viking, 1990.
Shuttleworth, Sally, and J. R. R. Christie, eds. *Nature Transfigured: Science and Literature, 1700–1900.* Manchester: Manchester UP, 1989.
Todd, Janet. *A Biographical Dictionary of British and American Women Writers, 1660–1800.* Totowa: Rowman and Allanheld, 1984.
Trimmer, Sarah. *An Easy Introduction to the Knowledge of Nature, and Reading the Holy Scriptures, Adapted to the Capacities of Children.* London, 1780.
Trimmer, Sarah. *Fabulous Histories, Designed for the Instruction of Children, Respecting Their Treatment of Animals.* London, 1796.
Twining, Elizabeth. *Illustrations of the Natural Orders of Plants, with Groups and Descriptions.* London: Joseph Cundall, 1849–55.
Wakefield, Priscilla. *Domestic Recreation; or, Dialogues Illustrative of Natural and Scientific Subjects.* Philadelphia: Robert Carr, 1805.
Wakefield, Priscilla. *An Introduction to Botany, in a Series of Familiar Letters.* London, 1796.
Wakefield, Priscilla. *Mental Improvement; or, The Beauties and Wonders of Nature and Art* (1794–97). Ed. Ann B. Shteir. East Lansing, MI: Colleagues P, 1995.
Wood, J. G. *The Boy's Own Book of Natural History.* London: George Routledge and Sons, 1883.
Zornlin, Rosina M. *Recreations in Geology.* London: Parker, 1839.
Zornlin, Rosina M. *Recreations in Physical Geography; or, Earth As It Is.* London: John W. Parker and Son, 1855.

PART 2

RECUPERATING THE WOMEN

2

The Invisible Woman

Stephen Jay Gould

Following Khrushchev's revelations of Stalin's less than saintly persona and procedures, the Soviet Union revised its official version of Communist Party history during the twentieth century. I bought a copy of this new edition and immediately turned to the index to learn the latest word on Uncle Joe. I found that he had suffered the worst of all fates: he simply wasn't there. And I thought to myself: love him or hate him, but how in hell can you tell the story of twentieth-century Russia without him? The keepers of official records had used the primary device of excommunicators, anathematizers, and ostracizers throughout history: there is a fate far worse than death or the rack, and its name is oblivion—not the acceptable fading of an honored life that passes from general memory as historical records degrade but the terror of unpersoning, of being present (either in life or immediate memory) but bypassed as though nonexistent.

Whole groups have suffered this fate as a consequence of general prejudice rather than special excoriation. As a primary example from my own field of evolutionary reconstruction, nearly all older theories for the "ascent of man" limited their concepts by the same prejudice that set their choice of words. Until the feminist movement provoked a salutary expansion to the entire human race, nearly all theories attributed our shared capacities for language, intelligence, and other valued properties of mind entirely to the activities of prehistoric males. Thus we learned that language arose from the coordination needed to hunt large animals (an all-male activity in conventional reconstructions) or that consciousness itself emerged from the more complex mental functioning required to stalk game (another male preserve). Women, under these theories, were simply invisible—sitting in the cave with the kids, I suppose (and so depicted in paintings and museum dioramas), but unmentioned in explicit text.

This ideological invisibility of prehistoric females was bolstered by a sexist sociology of disciplines that prevented living women from practicing the most prestigious parts of science—research and publishing. Only in this generation have women been entering science in substantial numbers. (I am proud that my own lab has included 50 percent or more of women graduate students during the past decade but I must admit that the first woman teaching fellow in our largest general course for nonscientists did not obtain her position until the early 1970s; she is now a distinguished researcher at the Smithsonian Institution.)

If intellectual women have been so restricted in our own day, consider the even greater limits imposed during the nineteenth century, the subject of this essay. In England, women were excluded from most major scientific organizations. The Geological Society of London did not admit women until 1904 (T. H. Huxley, to his discredit, had strongly supported the ban), the Linnean Society until 1919. Women fared better in botany, a subject considered suitable to the tastes and sensibilities of a "weaker sex." But even here, the reasons for limited acceptability took root in discrimination and fostered no egalitarian flowering. In an admirable study of "The Women Members of the Botanical Society of London, 1836–1856" (*British Journal for the History of Science*, 1980), D. E. Allen wrote:

> Botany could break the rules because it had the great good luck to be in keeping with both of the contemporary alternative ideals of femininity. On the one hand it was able to masquerade as an elegant accomplishment and so found favor with the inheritors of the essentially aristocratic "blue-stocking" creed, with its studied cultivation of an un-intense intellectualism. On the other, it passed as acceptable in those far more numerous middle-class circles which subscribed to the new cant of sentimentalized womanhood: the "perfect lady" of a repressive Evangelism.

Even so, women played only a subsidiary role when admitted. The Botanical Society of London began in 1836 with some 10 percent female membership, but the proportion dropped and remained at about 5 percent during the society's twenty-year life. Only one woman ever contributed a paper to the society's meetings, although she did not read the work herself but enlisted a male member as a surrogate for the occasion. No woman was ever elected to the council or served as an officer of the society. Women members could vote at meetings, but only (as the rules stated) after "having previously informed the secretary in writing of their appointment of some gentleman, being a Member of the Society, as their proxy for the occasion." Finally, the Botanical Society was, itself, an iconoclastic organization, and the more established scientific institutions continued their total ban. Allen writes:

> The Botanical Society . . . was one of that tangle of minor bodies which had gradually been springing up to cater for the large under-class of the scientifically inclined who, even if they had the intellectual attainments, could on social grounds scarcely hope for election to the major societies. In short, it was an organization of outsiders. And like its fellows, it reflected this in a self-consciously liberal stance that verged on radicalism.

Women with scientific interests were therefore confined to a narrow range of marginal activities, away from (or, at best, auxiliary to) the centers of prestige and innovation in research and publishing. Women could illustrate works written by men. The plates for John Gould's *Birds of Europe,* second only to Audubon in desirability and cost for modern book collectors, were drawn largely by his wife, who therefore deserves most credit for the work's reputation—a consequence of the figures, not the text. Incidentally, many of the other plates were done by Edward Lear, one of Europe's best scientific illustrators by profession, but better known to us today for his nonsense verse. Women could be collectors of specimens that were then turned over to men for formal description and publication. The beginnings of British vertebrate paleontology in the early nineteenth century owe more to the premier collector of this or any other age, Mary Anning of Lyme Regis, than to Buckland or Conybeare or Hawkins or Owen or any of the men who then wrote about her ichthyosaurs and plesiosaurs. The greatest collector of marine algae, Mrs. A W. Griffiths of Torquay, was warmly praised by the man who then wrote the most popular work on seaside botany: "She is worth ten thousand other collectors; she is a trump." Charles Kingsley stated that British marine botany would scarcely exist without her and, in a revealing choice of words, lauded her "masculine powers of research." Yet, as Lynn Barber writes in her excellent account of British popular science (*The Heyday of Natural History, 1820–1870,* Doubleday, 1980): "One genus and several species of seaweed were named after her, and she is mentioned with respect—almost with awe—by every Victorian writer on seaweeds, but she published nothing in her own name, and now survives only as an acknowledgment in other people's prefaces."

But the most public pathway for women lay in the writing of popular works in natural history—but only of a definite and characteristic genre: the saccharine and sentimental exaltation of nature's objects as illustrations of divine goodness and as guides to human reverence and proper behavior. Scores of women wrote an astonishing variety of such books, now almost totally forgotten, but then a conspicuous and profitable staple of publishing. These works are easily dismissed, as even Lynn Barber did in a sharp passage within a chapter devoted to praising those of her gender who persevered in science despite the obstacles:

> Unnumbered tribes of Victorian ladies seem to have written without ever doing an iota of research. Ladies were next in line to clergymen as relentless producers of popular natural history books, able at the rustle of a publisher's contract to launch into endless stories about . . . faithful dogs who rescued their masters from everything under the sun, and elephants Who Never Forgot. Most of their effusions were directed to other ladies, or to children (it is often hard to tell which) and characterized by glutinous sentimentality . . . and an ability to drop into verse at the least provocation.

I do not dispute Barber's assessment, but I believe that we must take this genre more seriously for many reasons, ranging from the scholarly (for insights provided into the history of women's social and intellectual struggles) to the ethical (respect due to marginalized people who bridle and suffer under imposed limitations but who, even on pain of illustrating a stereotype, must find some expression for creative impulses—blacks who preferred opera but entered the only available world of popular music, Jews like my grandfather who wished to be artists but ended up as skilled garment makers). I am happy to note that several scholars, particularly in feminist circles and programs for women's studies, are now rediscovering the invisible women who wrote popular books in natural history during the nineteenth century. I do not have the knowledge or experience to make a professional contribution in this specialized area, but I wish to record a personal encounter.

I recently purchased, at the decidedly low-priced end of a catalog from Britain's major purveyor of antiquarian books in natural history, a copy of a quintessential representative of this genre—*The Conchologist's Companion*, 1834 edition, by one Mary Roberts. I had not heard of it or her, but I was intrigued because I am a conchologist (a student of mollusk shells) by specialty and because I wanted to learn about this once important but now invisible genre.

As I tried to discover Ms. Roberts, I quickly encountered the bugbear of all scholarly research into the activities of people declared peripheral by society's gatekeepers to intellectual prominence. Such people are nearly invisible today. Nobody wrote about them during their lives, and they never became subjects for later historians. We are often left with the meager evidence of parish birth records, publishers' accounts, and epitaphs.

Mary Roberts wrote a dozen books on natural history, some apparently quite popular in her day, but she wins no more than a column in any biographical source. She was born in London in 1788, the daughter of a Quaker merchant (a common religious affiliation for women writers in Victorian times). Her family moved to Gloucestershire in 1790, but she returned to London (and left the Quakers) after her father's death. She lived in Brompton Square, London, for the remainder of her life, never married, and died on 13 January, 1864. I have been able to find absolutely nothing else about her life. The most common observation, mentioned by all five of our sources, is her frequent confusion "with another Mary Roberts, 1763–1828 . . . who dedicated to Hannah More an ambitious collection of poems." Such are the inevitable fruits of invisibility (and a common name). The biographical sources also say almost nothing about the content of her work, beyond listing titles—although this evidence is potentially available (if also hard to get, for few libraries maintain collections in popular writing of past ages). My oldest source, *A Critical Dictionary of English Literature*, from 1870, refers to Mary Roberts as "a useful and popular English authoress." My latest, *The Feminist Companion to Literature in English*, from 1990, notes that "even when writing for young children, she carefully names her sources." She wrote some nonscientific books, most notably a compendium on lives of famous women entitled *Select Female Biography* (1821), and an 1823 work with the intriguing title: *Sequel to an Unfinished Manuscript of H. Kirke White's to illustrate the Contrast Between the Christian's and the Infidel's Close of Life*. But most prominent are her dozen or so works in popular natural history, including (in chronological order): *The Wonders of the Vegetable Kingdom Displayed; Annals of My Village: Being a Calendar of Nature for Every Month in the Year; Domesticated Animals Considered with Reference to Civilization and the Arts; Sister Mary's Tales in Natural History; The Seaside Companion; Wild Animals; Sketches of the Animal and Vegetable Productions of America; Ruins and Old Trees Associated With Memorable Events in English History; Flowers of the Matin and Evensong, or Thoughts for Those Who Rise Early, in Prose and Poetry; A Popular History of the Mollusca;* and (my favorite title) *Voices from the Woodlands, Descriptive of Forest-trees, Ferns, Mosses and Lichens.*

I cannot deny that *The Conchologist's Companion* is as conventional as conventional could possible be—as conservative, as mainstream, as establishment, as comfortable to all expectations, both of general argument and of particular style, expected from woman authors. (I do sense the contradiction implied by stating that work in a marginalized genre can be mainstream in content, but ideology and respect are different phenomena, and we all know only too well the phenomenon of the slave who lustily mimics his master's voice.)

The conventionality of argument helps us to grasp a superseded conceptual world in which God made all of nature in beauty and harmony, as an illustration of his power and goodness, and to provide the crown of his creation (us, of course) with a bounty of food, fuel, clothing, gems, and building material. A documentation of this system also helps us to understand, in an immediate and visceral way, the depth of the intellectual revolution promulgated by Darwin and his theory of evolution by natural selection.

Nature, to Mary Roberts, is pervaded by divine purpose; every organism works in harmony with all others toward the general good:

> All the various parts of nature are beautifully designed to act in concert. We see the hand of God employed in forming the lowest, and frequently, in our opinion, the most despicable creatures; assigning to each its station, and so admirably adjusting the mighty whole, that every particle of matter, and every living thing that creeps, or moves, upon the surface of the earth, is formed in subserviency to the general good.

God is so attentive that he even created some animals as food for others, being careful to locate them in accessible places. Mollusks exist, in part, to feed higher creatures:

> Some inhabit ditches and stagnant waters, where they afford a constant supply of food to such birds as frequent their banks; others, no doubt with the same benevolent design, incrust marine plants in sandy barren places, near the sea; a large proportion remain concealed in the deep recesses of the ocean, where they furnish food to the finny tribes; others adhere to floating sea-weeds, and abundantly supply the wants of marine birds; and, lastly, exotic snails abound in many uncultivated regions of the globe, where they frequently afford a welcome repast to the fainting traveler.

Roberts illustrates the principle of universal goodness and teleology by the standard device of taking an apparently harmful creature and demonstrating its actual contribution to the general good (and specific human benefit). Teredos, or woodboring clams (commonly called shipworms), seem to be noxious in their destruction of ships, piers, and pilings. But look again to a wider beneficence. First of all, they bore very carefully, in a mode divinely calculated to cause minimal harm (I confess that for all my vaunted desire to approach the past in its own terms, without ridicule born of current knowledge, this passage did make me howl): "But mark the protecting care of Providence. The destructive operations of these insidious animals are in a great degree obviated, by the singular fact of their generally perforating the wood in the direction of the grain."

Moreover, by reducing logs and clumps of vegetation to sawdust, teredos prevent the clogging of rivers and flooding of land. Finally (and here I howled even more), teredos "open a source of considerable riches to the inhabitants of Sweden" by "employing the vigilance of the Dutch." For the teredos, you see, force the Dutch to keep their dikes and ships in continual repair, thus requiring "a perpetual demand for oak, pitch, and fir," largely supplied by Sweden—and so "these apparently pernicious insects are continually at work at Amsterdam, for the advantage of Stockholm." Roberts therefore concludes:

> Cease then, my friend, to regard this creature as decidedly obnoxious. . . . The Creator has assigned it an important station among his works. The evil which it produces, is readily obviated by a little care and contrivance; but the good which it is appointed to effect, is incalculably great in the mighty scale of universal nature.

All aspects of nature illustrate and glorify God, even those features that seem to contradict each other. We may think that God made the pretty colors of shells for his (or our) delight, but they actually produce cryptic patterns that camouflage mollusks from their enemies:

> But why, illustrious naturalist, did your observations extend no further? Saw you nothing in these . . . shells, but an arrangement of colors to please the eye. . . . Saw you not, that the Almighty Creator of the universe, without whose permission a single hair does not fall from our heads, nor a sparrow from heaven to the ground, nor a shell, nor a pebble, is tossed with the billows on the shore, by investing them in these simple colors . . . provides against their utter extinction, by the depredations of sea-birds and rapacious fishes.

But in other passages we learn that God did make colors only for beauty, and that, as such, they stand in contrast to the utility of shell shapes and sizes. Roberts argues for adaptation in the shape and ornament of the pholad (rock-boring clam) shell, but not in the color:

> An ovate or oblong form is consequently the very best that could be adopted; and, moreover, the points with which it is covered and adorned, are evidently

designed to protect the shell from external injury. . . . At the same time a beautiful variety of tints evince that minute attention to the finishing and decorating of his works which the Deity so continually displays.

Finally, nature is not only well designed and full of beauty; she is also replete with moral messages for human betterment. The metamorphosis of butterflies symbolizes the liberation of the soul from an earthly body, while the unfolding of a flower represents our hope for mental enlightenment:

The botanist confesses, in the unfolding of the calyx . . . an attractive emblem of the expanding of the human mind, as it emerges from a state of ignorance . . . or in the gradual development of a plant, the progressive advancement of every moral excellence.

Roberts's conventionality of argument is fully matched by her faithful adherence to the style of presentation expected from women authors in this genre. When, for example, she discusses the utility of mollusks in human life, she emphasizes supposedly "feminine" themes of adornment, rather than "masculine" motifs of immediate sustenance. Her longest chapters treat pearls and purple dyes (classically extracted from snails), but she barely mentions the fact that many people eat clams and snails.

More important, and now getting closer to the bone of our urge to render judgment, even across centuries, Roberts invokes the putative hierarchy and stability of nature to argue that each of us must accept our appointed role in human society, even if we be placed at a severe disadvantage by accident of birth as a female or as a member of the working classes. Our ultimate reward, after all, is not in this world. Just as God has designed each species of mollusk for its appropriate environment, he has "assigned to every individual being, his respective sphere of action; and happy will it be for us, if we as steadily perform our portion of allotted duty, as these feeble creatures fulfill the purposes for which they are designed, in accordance with their respective instincts."

Later, in the book's most uncomfortable passage (for me at least), Mary Roberts echoes Alexander Pope in the *Essay on Man* by arguing that any change in appointed ranks will cause the exquisitely balanced apparatus of nature to tumble:

To this splendid superstructure, nothing can be added; neither can any thing be taken from it, without producing a chasm in creation, which, however imperceptible to us, would materially affect the general harmony of nature. All things were made by him, and without him cannot any thing subsist; besides, it seems as if he designed to teach us by the admirable arrangement of his creatures, that the different gradations in society are designed by his providence, and appointed for our good.

These overtly sexist and politically conservative themes lie exposed at the surface of *The Conchologist's Companion*. But I was struck even more by the pervasive "deep sexism" of Roberts's obedience to what her society viewed (and we, in large and unfortunate measure, continue to regard) as contrasting ideals of the abstract and eternal masculine and feminine—the surest guide, I believe, to a true understanding of *The Conchologist's Companion*, and other works of this once influential genre. I suspect that we can best grasp this vital, but largely covert, theme by returning to one of the most important essays in English letters—a document that many of us read (at least in excerpt) in Philosophy I, but have probably never thought about since: Edmund Burke's *Philosophical Enquiry into the Origin of Our Ideas of the Sublime and Beautiful*, first published in 1757.

Burke argues that our aesthetic senses are moved by two separate configurations, which he calls the sublime and the beautiful. These are truly distinct, for one is not the negation or the reciprocal of the other. The sublime (which Burke also calls the "great") is based on our instinct for self-preservation and founded in terror. Among its themes lie vastness, darkness, verticality, massiveness, roughness, infinity, solidity, and mystery. The beautiful, on the other hand, is rooted in pleasure and linked to our instinct for generation (necessary for the preservation of our race, but not so elemental as self-preservation and the sublime). The themes of beauty include smallness, smoothness, variety in shape but only by rounded rather than angular transitions, delicacy, transparency, lack of ambiguity, weakness, and bright colors.

Burke does not dwell on the correlations of these themes with conventionally sexist views of the masculine (sublime) and feminine (beautiful), but this contrast sets the basis of "deep sexism." Burke argues, for example, that women instinctively recognize the necessary link between beauty and weakness: "Women are very sensible of this; for which reason, they learn to lisp, to totter in their walk, to counterfeit weakness, and even sickness. In all they are guided by nature."

He also attributes the supposed grandness of male thought, and the timidity of female mentality, to this distinction: "The sublime . . . always dwells on great objects, and terrible; [beauty] on small ones, and pleasing. . . . The beauty of women is considerably owing to their weakness or delicacy, and is even enhanced by their timidity, a quality of mind analogous to it."

I submit that we cannot grasp the essence of Roberts's book, and the genre she represents, until we assimilate this classic distinction of the sublime and the beautiful. We must, above all, recognize that Roberts and her colleagues of like gender accepted this scheme and sought to be completely beautiful and not at all sublime—that is, essentially feminine by their light. (I also urge readers to acknowledge this distinction as both productive and supportive of the worst aspects of sexism—and to remember that liberation presupposes knowledge of the causes of oppression.)

Burke's criteria of beauty supply a key that opens *The Conchologist's Companion* to our understanding (rather than our ridicule based largely on puzzlement). The conceptual themes are all present: conventionality, timidity, boundedness, lack of surprise, rounded transitions. Even the physical appearances proclaim beauty rather than sublimity. These books by women tended to be small in size—printed at dimensions that publishers call duodecimo or small octavo, rather than the large octavo or quarto favored for books by male authors. Type sizes are generally small, and the engraved illustrations particularly delicate. The prose itself proclaims saccharine sentimentality, rather than raw power—particularly in the dum-de-dum of doggerel verses:

> Oh ! Who that has an eye to see,—
> A heart to feel,—a tongue to bless,
> Can ever undelighted be
> By Nature's magic loveliness!

The choice of subject matter—small and humble mollusks—matches the attributes of a female writer, as Roberts mentions again and again:

> In the prouder forms of animated being, in the towering cedar, or the turret-bearing elephant, nature appears to act in a manner analogous to the grandeur of her designs; whereas these feeble creatures are often passed by, as undeserving the attention of the naturalist; and yet what tokens of beneficence and power, what exquisite perfection is discoverable!

Even the words of praise, used by men in their favorable reviews of women's work, invoked the standards of circumscribed beauty, rather than awesome sublimity. The *Athenaeum* (journal of the intellectual men's club that included Darwin and Huxley as members) praised Mary Roberts's second book on conchology as "a useful and entertaining volume" (issue of 22 November 1851, page 1124).

I was going to end this essay here, with some words of exculpation for Ms. Roberts, acknowledging her utter submission to conventional expectations, but refusing to judge her too harshly—for the urge to create can be so overpowering, and the pain of self-imposed silence so overwhelming, that we sometimes kowtow to the most iniquitous of limitations. (I dare not, as a white man, criticize Stepin Fetchit or Mantan Moreland for playing the only, if degrading, roles that Hollywood then allowed black actors. And I will not castigate any woman who needed to write, but could be published only by adhering to standards of Burkean beauty.)

But nothing in our complex world ever ends so cleanly. As I reread and considered, I began to see more in Ms. Roberts. I began to pick up an undertone of rebellion—muted to be sure, but unmistakably present. I began to realize that Mary Roberts had not totally internalized the norms imposed upon her, and that some flicker of feminine anger smoldered on the pages of her small and beautiful text.

One passage particularly caught my attention. Ms. Roberts often cites the conventional theme that nature will always hide secrets from human probing, and that we should not be too arrogant in our claims for understanding. She usually takes the standard line of attributing this prevailing mystery to masculine power—that is, to God's omniscience as creator of all nature versus our mental midgetry. (She writes, for example: "In many instances we are unable to comprehend the intentions of the Deity with regard to the construction of his creatures.") But in one striking passage, she identifies the cause of obstruction as *feminine* nature—and she explicitly contrasts the necessary victory of this female power with the conventional image (used explicitly by Bacon and later writers) of science as male and active, seizing (almost raping) knowledge from passive, feminine nature:

> It seems as if maternal nature delighted to baffle the wisdom of her sons; and to say to the proud assertors of the sufficiency of human reason for comprehending the mysteries of creation and of Providence, "Thus far can you go, and no further"; even in the formation of a shell, or its insignificant inhabitant, your arrogant pretensions are completely humbled.

I knew that I had to probe further into Mary Roberts's hidden motives and feelings. So I located her paleontological book, *The Progress of Creation* (1846 edition), in the stacks of Widener Library (actually not in the stacks, but further into purgatory at the book depository for rarely used volumes and requiring a two-day wait for delivery—another sign of invisibility for this genre).

I was struck by an immense difference in style within a basic similarity of drearily conservative content. *The Conchologist's Companion* may have made me sad for its bowing to limitations of a genre, but *The*

Progress of Creation made me mad for its ignorant pugnaciousness in clinging to clearly disproved religious formulations. In her geological book, Mary Roberts takes a hard creationist line in insisting upon biblical literalism—with creation in six days of twenty-four hours and a total age of only a few thousand years for the earth. She states in no uncertain terms:

> Throughout this volume, I have ever kept in view, that the heavens, and earth were finished, and all the host of them in six days; and that no theory, however plausible, can be admitted in opposition to the Divine Record.

She uses the writings of catastrophist geologists to assert the reality of Noah's flood and to argue that this deluge created the entire geological record in one grand gulp. But her argument is either ignorant or disingenuous. By 1846, all serious catastrophists, including all the men falsely quoted by Mary Roberts, knew that any flood within human history could only represent the latest catastrophe in a long series of earlier crises stretching throughout an immensity of geological time.

Yet Mary Roberts wades in, dukes held high and swinging at all the greatest male scientists of Europe. She thinks that mastodons are carnivorous and lambastes Georges Cuvier, the Newton of natural history, for his (evidently correct) belief that these elephants ate plants: "The carnivorous elephant, or Mastodon of the Ohio, is one of the most remarkable. Cuvier describes this animal as herbivorous, but surely without reason." This is absolute nonsense from Ms. Roberts's pen—but it is sublime nonsense.

What was Mary Roberts really thinking when she wrote *The Conchologist's Companion* by the rules of sweet, delicate, inoffensive prose—when we know that she could also write in pugnacious sublimity? Were her personal views on obedience and natural harmony as conventional as those she presented? Did she accept the limitations placed upon women, or did she seethe inside yet keep her own counsel? I doubt that the records exist to answer such questions—for a conventional history told largely by males did succeed in rendering her nearly invisible.

I wonder. What really went on behind the mask of acceptance and convention respected by most women writers on natural history? Perhaps we should invoke one of the great women of strength from our literature—Little Buttercup of Gilbert and Sullivan's *HMS Pinafore*. She tries to tell the captain that "things are seldom what they seem"—perhaps also that women often hide the pain and anger of ages under a soft surface of acceptance:

Though to catch my drift he's striving,
I'll dissemble—I'll dissemble;
When he sees at what I'm driving,
Let him tremble—let him tremble![1]

Mary Roberts (1788–1864): Major Works

The Wonders of the Vegetable Kingdom Displayed (1822)
The Conchologist's Companion (1824)
Annals of My Village: Being a Calendar of Nature for Every Month in the Year (1831)
Domesticated Animals Considered with Reference to Civilization and the Arts (1833)
Sister Mary's Tales in Natural History (1834)
Sketches of the Animal and Vegetable Productions of America (1839)
Voices from the Woodlands, Descriptive of Forest-Trees, Ferns, Mosses, and Lichens (1850)
A Popular History of the Mollusca (1851)

Notes

1. People often ask how or why I keep turning out these monthly essays after nearly twenty years without a rest. The answer is so simple: Burke's essay is one of the greatest documents ever written in our language—for all that I find repulsive therein (amidst other insights that are truly wonderful). My volume of Burke has always stood accessible on my bookshelf—as part of volume 24 in my set of Harvard Classics. My grandfather bought this set in the 1920s, when he decided that he had to provide for himself and his children an opportunity for the education that he had never received. I doubt that he ever consulted more than a volume or two. My mother read and reread the volume on Aesop, Grimm, and Anderson during her childhood. I shot arrows at the set during my youth—and my volume 24 is transfixed with nine holes marking my successes. I always like this particular volume because the spine says "On the sublime French revolution." I never would have so characterized this episode in history, so the title always puzzled me. I don't know when I ever thought of actually opening the book, but I do remember my delight in discovering that the title mixed two separate works—Burke's essay on the sublime and his reflections on the revolution! These generational and ontogenetic continuities give meaning and structure to our lives. Then I shot arrows; now I want to read. I could have pulled this book off the shelf and read Burke's essay anytime—but I never did and, to be honest, I doubt that I ever would have done so amidst all the crazy and maniacal crowded business of our lives. My decision to write this essay on Victorian natural history brought me to Burke, and I had the wonderful privilege of reading this great and influential document for pleasure, and with some maturity of judgment, rather than in the rush of a course and for a grade at too young an age. For this privilege, I am most profoundly grateful. If I didn't write these monthly essays, Burke would probably have stayed on my shelf until the day I died.

PART 3

DISSEMINATING KNOWLEDGE: ENGLAND

3

Fictionality, Demonstration, and a Forum for Popular Science

Jane Marcet's *Conversations on Chemistry*

Greg Myers

Scientists and others often lament the split between science and general culture. Though science is presented in many popular forms—movies, ads, television game shows, toys—scientists acknowledge that the public can not be expected to understand much of what is going on. Historians of science and literary historians point out that some classics of science, such as Lyell's *Principles of Geology,* could be read and discussed by the general educated upper-class public of their time. But at some point in the history of every science, the audience for important publications became limited and the forms became specialized in scientific articles. While the development of specialist genres has been well studied (Bazerman 1988; Dear 1991; Selzer 1993), there has been less attention to the other side of the split, the genres that developed to present science to the public, including textbooks, magazines, lectures, and popular treatises. Popularizations are still too often treated as a defective translation of a primary text. I would argue that it is more productive to take popularizations in their own terms, and to look at what nonscientists do with science. When we stop focusing on the split between scientists and the public, we see that there is no single public for science. We should look instead for the ways various splits are worked out: between masculine and feminine, mass and elite, young and old, metropolitan and provincial.

Figure 3.1. Jane Haldimand Marcet (1769–1858). Photograph courtesy of the Chemical Heritage Foundation.

I will compare a short passage from the work of perhaps the greatest popularizer of the first half of the nineteenth century, Jane Marcet (1769–1858), to two other treatments of the same discovery. Many nineteenth-century literary and scientific figures attested to Marcet's influence (see her entry in the *Dictionary of National Biography*); her *Con-*

versations on Chemistry served as one of the first textbooks on the subject, and her *Conversations on Political Economy* was the most widely read work on a field that was then at the center of intellectual debates. Unlike her contemporary (and friend) Mary Somerville, she did not see herself as contributing to the fields about which she wrote; she saw herself as a popularizer of the work of others. But, as we will see, the modesty of this role is complex and ambivalent. Marcet had inherited a large fortune, and married a doctor who was best known for his analysis of spa water and who later lectured in chemistry, so she did not (like her contemporary and friend Harriet Martineau) write for money. She published *Conversations on Chemistry* when she was in her late thirties, and for the rest of her long life she published new works or revised editions nearly every year. By the sixteenth edition of *Conversations on Chemistry* (1853), it had become rather dated, and had lost the excitement of contact with the latest discoveries that had filled the earlier editions. But by then it (and the other *Conversations*) had reached the libraries of schools, colleges, and mechanics' institutes throughout Britain and North America (Lindee 1991). Marcet was writing at a key moment in the popularization of chemistry, as textbooks proliferated and courses began, and the subject of chemistry began to reach beyond professionals and industrialists to wider publics.

I will compare the description of Davy's discovery of potassium (reprinted in an 1813 pirate American edition of the third edition of *Conversations on Chemistry*), to two other treatments, a lecture by Davy himself, and a review of the lecture by the influential politician and founder of the Society for the Diffusion of Useful Knowledge, Henry Brougham. At first glance, these versions of the discovery are strikingly similar, so that we might think that both Brougham and later Marcet just paraphrased Davy's text. But if we look more closely at the passages, we see that Brougham and Marcet interpreted the discovery and projected an audience for it, taking it in opposite directions. Marcet did not just explain the work of Davy; she created a forum in which it could have cultural effects.

My analysis here concerns the relation of her form to her forum. The fictional form of *Conversations on Chemistry* (as of Marcet's other *Conversations* on botany, natural philosophy, and political economy) involved a woman, Mrs. B., discussing topics with two girls. The dialogue form has a long history as a form for presenting original scientific work (Myers, "Science," "Fictions"). Marcet and contemporaries like Samuel Parkes, Jeremiah Joyce, Maria Jacson, and Priscilla Wakefield used it specifically to popularize work already published in other forms (Benjamin; Henderson; Knight, "Accomplishment"; Shteir, "Botany," "Botanical Dialogues"). The forum was apparently that of a home like that in the story, where science is part of a program of self-improvement for the daughters of the middle classes. Of course the effects of *Conversations* went far beyond this named audience: the book was read by college students and by working-class self-improvers, most famously Michael Faraday, who was led by it to Humphry Davy and his own career in science. But its wider and more complex effects may have depended on its careful definition of its place, not defying norms but using conventionality to clear a space for science in the domestic world.

Once we deal with various forums, and not with "the public," we can see how popular science was, from the outset, gendered. What place did these contemporaries assign to Jane Marcet and her contemporaries? It was, most obviously, a matter of *audience:* women write for children and for other women (even Somerville's very advanced *On the Connexion of the Physical Sciences* was dedicated to the Queen and nominally addressed to women). It was also a restriction of *fields.* Botany is female, but physics is not; philanthropy is female, models of political economy are not. It could be a matter of *practices:* women observe, while men experiment. Or it could be a matter of form: men write treatises and give lectures and do demonstrations; women write conversations, letters, and tales. In terms of these oppositions, Jane Marcet is apparently rather conventional. But by writing a fiction, about observation, for women and children, she could create a forum in which women could do chemistry.

The fictional frame of *Conversations on Chemistry* is usually seen now as an attempt to make science attractive to an audience who might otherwise find it boring, as in this comment by the historian David Knight: "The style of the letter, the conversation, the catechism, or the lecture appealed to those popularizing science and afraid that the neophyte might be put off for life by too dry a style" (*Age of Science* 136). This view takes the line between fact and fiction for granted, and takes the act of fictionalizing as the taming of a primary factual text, making it more easily approachable. But the fictionality of such works raises interpretive problems. When we interpret a text, we construct a world of referents around it. In *Conversations on Chemistry,* some of these referents are interpreted as part of the reader's actual world, such as Humphry Davy, or potash, or the equipment for collecting a gas, while others are taken as part of a world constructed in the book, such as Mrs. B., the house, or Caroline's dress. In Dickens's *Bleak House* (1853), all the characters are fictional, but the narrator insists on the spontaneous combustion of one character as a physical possibility. In Beatrix Potter's *Tale of Mrs. Tiggywinkle* (1905), the girl and Littletown and the types of animals are all said to be actual, but the behavior of the animals is fictional—the hedgehog talks and washes clothes. In Joan Aikin's *Black Hearts at Battersea* (1965), the setting is recognizable as nineteenth-century England, but it is an England loyal to James the Third, after a successful reestablishment of the Stuarts, and threatened by the rebellious Hanoverians. In Ursula Le Guin's *Always Coming Home* (1985), the scholarly forms are familiar and the place is recognizable, but the time is not continuous with ours, as the first sentence announces: "The people in this book might be going to have lived a long time from now in Northern California." We need a theory of fiction that can account for all these kinds of fictionalization, for they occur not just in the wilder reaches of science fiction and postmodernist metafictions, but in children's books, advertisements, political cartoons. Such a theory would treat fictional conversations like Marcet's as part of a broader continuum of fictional texts, and not just as a rather quaint pedagogical subgenre that flourished only in the early nineteenth century. As Thomas Pavel and Marie-Laure Ryan have shown, a fiction raises questions about how we map the world of the text onto our world, and thus about what we take for granted as surrounding a discourse. In this case, it raises questions about how a text embodies the forum in which it is delivered, and how it moves beyond that forum.

It helps to understand the world of Marcet's *Conversations* if we consider other textual representations of forums for chemistry. Let us start where Marcet started, with Humphry Davy's lectures at the Royal Institution, from 1801 to 1812. Fortunately, these lectures have been extensively studied by Jan Golinski, who sees them as representative of some key transformations in the science of the time: (1) a career strategy (Davy's creation as a genius figure); (2) a disciplinary strategy (the advancement of chemistry as practically applicable); (3) an institutional strategy (finding a fashionable audience for an institution founded for the working class); (4) a methodological strategy (showing the need for expensive equipment like the voltaic pile); and (5) a political strategy (aligning chemistry with the metropolitan elite rather than with provincial radicals). The defining characteristic of the Royal Institution Lectures (still visible in the annual Christmas Lectures broadcast by the BBC) is demonstration, the public performance of processes that would make phenomena visible to a wide audience. Golinski shows how the dramatic effects demonstrated in these lectures were crucial to their popular success, and how the demonstrations provided a basis for Davy's triumph in his interpretation of oxymuriatic acid.

It is striking that many of the contemporary accounts of these lectures describe them in terms of gender. The Earl of Minto found in the audience, not only a selection of the aristocracy and prominent politicians, but also "there are a great many women, principally matrons with young daughters, who take notes" (quoted in Foote 10). The poet Southey divided the audience into two parts: "Part of the men were taking snuff to keep their eyes open, others more honestly asleep, while ladies were all upon the watch, and some score of them had their tablets and pencils, busily noting down what they heard, as topics for the next conversation party" (quoted in Foote 11). A French traveler, Louis Simond, made a similar comment on bored men and note-taking women. "Women alone consider themselves neither above nor below Mr. Pond or Mr. Davy. In fact, public lectures are only useful to those who know little, and aspire to little" (quoted in Foote 11).

Simond's account suggests why the lectures—and popular science in general—posed a dilemma for gentlemen. Davy presented a figure that stood outside systems of status; the only options were the boredom of those that denied that chemistry counted as knowledge, or the condescension of those who claimed they already knew it. Women did not face this dilemma; they could enjoy the displays as part of the vague self-improvement prescribed for daughters of the upper classes who had no educational institutions. But their interest too was embarrassing. Why were male observers so disturbed by their taking notes? (Of course one can imagine equally scathing comments if they had not taken notes.) The treatment of science as a public display was one thing, but the removal of this display to another forum, that of the drawing room and "conversation," was seen as trivializing. Note taking was both passive submission to this man, which the men avoided with their boredom, and the active appropriation of his knowledge for feminine purposes. So it is important to see how Marcet handled these demonstrations in her *Conversations*.

The Bakerian Lectures were a prestigious annual series at the Royal Society; they were presented by invitation, published in the *Philosophical Transactions*, reviewed in the quarterlies, and later pirated to a much wider audience (for instance, I found an abridged version as an appendix to an American edition of *Conversations on Chemistry*). In presenting them, Davy was moving from the Royal Institution and its fashionable audience to the Royal Society and the metropolitan scientific establishment. When Davy summarized his recent researches for the Royal Society, he did not draw on the spectacular effects used in his public lectures, but they are referred to in the texts. We can see how Davy maintains the factuality of his account in his use of tense in setting up events, his attribution of agency to humans or equipment, his use of adjectives describing the effect on an audience, and his anticipation of objections. It is a situation for telling based on a rhetoric of showing. Let us take one set of experiments, the decomposition of potash that enabled Davy to identify potassium as an element. Davy has already described how a huge voltaic cell, connected to a solution of potash, made for a spectacular flash but only decomposed the water, not the substance under examination. He continues:

I tried several experiments on the electrization of potash, rendered fluid by heat, with the hopes of being able to collect the combustible matter, but without success; and I only attained my object, by employing electricity, as the common agent for fusion and decomposition. Though potash, perfectly dried by ignition, is a non-conductor, yet it is rendered a conductor by a very slight addition of moisture, which does not perceptibly destroy its aggregation; and in this state it readily fuses and decomposes by strong electrical powers.

A small piece of pure potash, which had been exposed for a few seconds to the atmosphere, so as to give conducting power to the surface, was placed upon an insulated disc of platina, connected with the negative side of the battery, of the power of two hundred and fifty of six and four, in a state of intense activity; and a platina wire communicating with it on the positive side, was brought in contact with the upper surface of the alkali. The whole apparatus was in the open atmosphere.

Under these circumstances, a vivid action was soon observed to take place. The potash began to fuse at both its points of electrization. There was a violent effervescence, at the upper surface: at the lower or negative surface, there was no liberation of elastic fluid; but small globules, having a high metalic lustre, and being precisely similar, in visible characters to quicksilver, appeared; some of which burnt with explosion, and bright flame, as soon as they were formed, others remained and were merely tarnished, and finally covered by a white film, which formed on their surfaces. These globules, numerous experiments soon showed to be the substance I was in search of, and a peculiar inflammable principle the basis of potash. I found that platina was in no way connected with the result, except as the medium of exhibiting the electrical powers of decomposition; and a substance of the same kind was produced, when pieces of copper, silver, gold, plumbago, or even charcoal were employed for completing the circuit. . . .

. . . After I detected the bases of the fixed alkalies, I had considerable difficulty to preserve and confine them so as to examine their properties, and submit them to experiments; for, like the alkahests imagined by the alchemists, they acted more or less upon every body to which they were exposed.

The fluid substance amongst all those I have tried, on which I find they have the least effect, is recently distilled naphtha. In this material, when excluded from the air, they remain for many days without considerably changing, and their physical properties may be easily examined in the atmosphere when they are covered with a thin film of it. . . . The action of the basis of potash on water exposed to the atmosphere is connected with some beautiful phenomena. When it is thrown upon water, or when it is brought into contact with a drop of water at common temperature, it decomposes it with great violence, an instantaneous explosion is produced with brilliant flames, and a solution of pure potash is the result. (338–39)

In many ways this passage is similar to any experimental account (Bazerman). It begins in first person, with the statement of his goals, but then it immediately shifts to passive voice. The *I* reenters at the end of the account, in the discussion of its interpretation. Despite the lack of equipment in this setting, demonstration plays a key role. Davy gives a vivid description of perceptions: a *vivid* action, a *violent* effervescence, a *bright* flame, *beautiful* phenomena, *great violence,* and *brilliant* flames. He does not perform the demonstration, but does represent the visual effect textually. These passages contrast with the lack of visual description when he is describing apparatus or defining properties. The circumstantial details are not strictly necessary, but testify to the actuality of the effects. It is the literary technology that Shapin described in the work of Boyle as creating a sense of "virtual witnessing."

But there is also another sense of *demonstration* here, that of argument and proof. At the end of the quoted passage, Davy begins anticipating objections to the claim that he has decomposed potash; the substances formed could have come from the platinum disc. He then breaks with the chronological account, and compresses an array of separate experiments to confirm that the globules could not have come from the disc, and goes on to show they could not have come from the atmosphere to which the potash had been exposed. In these passages, the aim is no longer to give a vivid impression of natural phenomena, but to eliminate other interpretations of it. The reasoning is left implicit, and the audience must make any links itself. Consider the *and* in "and a substance of the same kind was produced, when pieces of copper, silver, gold, plumbago, or even charcoal were employed." The force here is that of *because,* but to read it this way we must already see the relevance of employing other substances. When charcoal is referred to as "or *even* charcoal," it is assumed we know why it would be the least likely of these substances to be used in producing this substance. The larger form of this section of the lecture is a movement from the event to the qualities of this new element, which again assumes an informed audience interested in the general framework of the discipline. The movement is marked by a shift of tense, from the past tense of "a small piece . . . was placed," to the present tense of "when it is thrown upon water." Though this description of potassium bursting into flames is as vivid as the earlier passage, it is here an example of the properties of the element, not a narrative account of a particular past event.

Davy's Bakerian Lecture is certainly not like a modern scientific article. It is more like some popularizations in its concern with visual effect. But it does not create a public forum. In fact, it moves from the more public forum of the Royal Institution to the elite of the Royal Society, where the audience is expected to be able to draw on a shared background in chemistry, to follow his reasoning and do without his spectacular shows.

Davy's Bakerian Lectures were regularly reviewed by the *Edinburgh Review,* which had just begun its long run of domination of Whig intellectual and literary circles. This in itself may seem surprising; today the *New York Review of Books,* which has a similar policy of pegging lengthy essays onto book reviews, does not pay much attention to *Nature* (Young). What is more surprising to the modern reader is that the account was indeed a critical review, giving the personal opinion of the anonymous author: "Mr. Davy's last course of experiments . . . showed more ingenuity and dexterity than the present" (Brougham 395). If the *New York Review of Books* did comment on *Nature,* one can hardly imagine them criticizing the detector design at CERN. The author (as identified in the *Wellesley Index to Victorian Periodicals*) was Henry Brougham, later a politician, a proponent of popular education through the Society for the Diffusion of Useful Knowledge, and a Lord Chancellor. He had taken chemistry at Edinburgh, so he could present himself as informed and critical. Thus he solved the problem of status that seems to have worried the men described by Simond; he took it for granted that he was a qualified referee. But Brougham's stance toward the lectures was not due only to his own rather unusual education. He could take it for granted that science was part of his sphere of political and social interests, as much as any other topic under review. The controversy was not, as it is today, about its accessibility to "the public." It was about the suitability of this knowledge for those outside the elite, the implications for the nation, for politics, and for society.

Brougham's account is in many ways a paraphrase of Davy's. But the account is transformed by the heavy addition of evaluative language. And this is made possible by what I take to be a fictional device, the anonymity of the reviewer. This anonymity, conventional to the reviews, made the author a fictional figure, not located exactly in the world of the readers. This would be true even where the identity of the writer was, as it must have been, widely known, for the anonymity allowed him (or in very rare cases her) to speak as a general voice of the

Edinburgh Review. The anonymity allows the personal figure of Davy to be evaluated by an impersonal figure of a certain intellectual circle, outside of scientific institutions.

Brougham's account of the experiments is subtly evaluative; it also establishes explicitly the kinds of logical links that were implicit in Davy's account:

> He then tried the alkalis fused by heat, and without success; for it seemed evident, that the fusion and action must come both at once from the electricity. Accordingly, having slightly moistened the surface of perfectly dry potash, so as to render it a conductor, he placed it on an insulated disc of platina, and placed the positive wire on the upper surface of the potash. A remarkable action now ensued: the salt fused at the wires, at the lower surface, without any effervescence; but, at the upper, with violent effervescence. At the lower surface, however, small globules like quicksilver were perceived to emerge as the process went on, and many of them burst with explosion and a bright flame; others, without explosion, soon became covered with a white crust on continuing exposed to the air. The same phenomena were produced, when, instead of platina, other metals as copper, gold, &c. were used, or plumbago, and even charcoal. The metallic globules, therefore, had nothing to do with the disc or wire employed: and the experiment was equally independent of the air—for it succeeded just as well in an exhausted receiver. . . .
>
> Such is the decisive and most satisfactory evidence by which it is ascertained that the fixed alkalis are compounds of oxygen and metallic bases, or that they are in truth metallic oxides. The metals are substances hitherto quite unknown to chemistry; and Mr Davy, as might easily be imagined, lost no time in examining their peculiar properties. It is unnecessary to detail the various experiments he made for this purpose. . . . (395–96)

For Brougham's imagined public the overall plan of the experiments is obvious ("as might easily be imagined"); thus he either assumes in them, or more likely, attributes to them, a detailed understanding of scientific practice. But for this audience "it is unnecessary to detail the various experiments". Nor are they concerned with the showy fireworks; it is the properties of the new element that are supposed to surprise them, its place in a larger system of knowledge. Brougham makes two main additions to Davy's text: evaluative words as in *success, remarkable* action, and *brilliant* discoveries, and logical links like *accordingly,* and *therefore.* Later, in an account of Davy's experiments on oxymuriatic acid, he allows himself to express skepticism about both the reasoning and the experimental setup.

Brougham is also evaluative in a later passage dealing with Davy's innovations in nomenclature (Marcet would also be crucially concerned with names). Because they have been accepted, the names he mocks no longer sound like jargon to us. "He names them Potassium and Sodium—names, as he remarks himself, more significant then elegant, but we are greatly relieved at finding them no worse." This naming becomes the occasion for some elaborate irony. Brougham says the rumor was that they would be labeled Sodagen and Potagen (by analogy to Lavoisier's Oxygen—perhaps too French), while others had suggested terms praising the monarchy and the church. This allows Brougham a dig at the Tories, seen to control the London scientific establishment. "We well knew no such thing was ever listened to by the discoverer himself, whose political sentiments are as free and manly as if he had never inhaled the atmosphere of the Royal Institution" (399). The *Edinburgh Review* brings Davy out of the uncomfortably ambiguous world of the public lecture into the more markedly manly world of politics, by setting an anonymous reviewer above him. The review is not just for the public, but for a special section of the public, a male audience as opposed to the female audience of the Royal Institution, an educated elite rather than a mass audience, a Whig review rather than a rather Tory (at that time) Royal Society. To be brought out into public is to be subjected to public scrutiny. The anonymous voice of the reviewer is offered to speak for this public.

Jane Marcet's use of fictional forms can be seen as a defensive move, avoiding just those larger claims made so strongly by Brougham. She could not hold forth on political connections, or evaluate Davy's place in the pantheon of science. But she could use fiction to bring Davy's science into a world in which she could control it, where she understood the audiences. She says in her "Preface" that the *Conversations on Chemistry* were specifically designed to prepare women for the "the excellent lectures delivered at the Royal Institution by the present Professor of Chemistry." In the first edition, she, like Brougham, was anonymous, so she could stand for all the note-taking women. After she had herself had some private tuition in chemistry with "a friend,"

> Every fact or experiment, attracted her attention, and served to explain some theory to which she was not a total stranger; and she had the gratification to find that the numerous and elegant illustrations, for which that school is so much distinguished, seldom failed to produce on her mind the effect for which they were intended. (iii)

There is no sense here of Marcet appropriating knowledge of chemistry for a new forum. The stated aim of the *Conversations* is merely to do some remedial work so that the lectures have their proper effect. And when she states her obligatory diffidence about authorship, the openness of the lectures (leaving aside men's annoyance at the note-taking women) serves as one of two lines of defense:

> she felt encouraged by the establishment of those public institutions, open to both sexes, for the dissemination of philosophical knowledge, which clearly prove, that the general opinion no longer excludes women from an acquaintance with the elements of science. (iv)

The other line of defense, though apparently modest, stakes an important claim. She was, she said,

> flattering herself, that whilst the impressions made upon her mind, by the wonders of Nature studied in this new point of view, were still fresh and strong, she might perhaps succeed the better in communicating to others the sentiments she herself experienced. (iv)

This amounts to a claim to superiority in another forum, to an understanding of the processes of learning that she herself had experienced, and of the learners. Unlike the uneasy men in Simond's account of Davy's lectures, she did not either have to know the material already, or to pretend boredom. Mrs. B., Caroline, and Emily could act out the processes of learning that were missing from either Davy's or Brougham's account.

The fiction of the *Conversations* delineates a new forum. The boundary drawing in the fiction can be seen in the way Mrs. B. presents demonstrations. The same issue is raised in Galileo's dialogues: historians argue about which episodes were meant as reports of actual experiments, and which were meant as thought experiments. Marcet's pedagogy depends on showing rather than just telling: "The dry white powder which you see in this vial is pure caustic potash" (152). The deictic *this* is the fictional element—it asks us to see the substance here and now, not just to recognize that it is, in general, a white powder. Marcet could, of course, dramatize all of Davy's experiments: if the fictional Mrs. B. can gesture to a fictional vial, then she could just as well gesture in fiction to, say, a huge voltaic cell with leads connected to a platinum disc. But she makes a point of excluding these larger demonstrations:

> I am sorry that I cannot show you the combustion of perfect metals by this process, but it requires a considerable Galvanic Battery. You will, however, see these experiments performed in the most perfect manner, when you attend the chemical lectures of the Royal Institution. (134)

At other points, Mrs. B. refers to demonstrations she cannot do because they require great experimental skill (216) or difficult-to-procure materials (225). She seems to put off the experiments with inhalation of nitrous oxide, which were such a sensation in Davy's earlier career, causing all sorts of bizarre behavior and perceptions, and in response, criticism of chemistry in general. She forbids the girls to try, for "the nerves are sometimes unpleasantly affected by it" (212), but "I daresay we shall find some member of the family who will be curious to make the experiment of respiring it" (213). But it takes time to collect the gas, and the experiment seems to be forgotten. Since there is no reason in the fiction why these demonstrations cannot be included, the distinction between possible and impossible demonstrations reinforces the differences between laboratory and home.

In naming, as with demonstrations, the effect is to establish a division between the home and the laboratory. The girls must observe a complex linguistic etiquette; while Brougham could be high-handed in evaluating new terms, they must be careful about sounding pedantic. Mrs. B is talking about the tin preventing iron from rusting.

Caroline: Say rather *oxydating*, Mrs. B.—Rust is a word that should be exploded in chemistry.
Mrs. B: Take care, however, not to introduce the word oxydate instead of rust, in general conversation; for either you will not be understood, or you will be laughed at for your conceit.
Caroline: I confess that my attention is, at present, so much engaged by chemistry, that it sometimes leads me into ridiculous observations. Every thing in nature I refer to chemistry, and have often been laughed at for my continual allusions to it.
Mrs. B: You must be more cautious and discreet in this respect, my dear, otherwise your enthusiasm, although proceeding from a sincere admiration of the science, will be attributed to pedantry. (145)

Battles over nomenclature characterize some phases of all disciplines (chemistry and geology then, linguistics and psychology today). But here the issue is one not of classification of things, but of distinction of forums—the issue is who can use what terms where.

The passage in which Mrs. B. discusses Davy's potassium experiments is typical in the way it begins with justifying the lesson to the two girls.

Caroline: But I have heard that these discoveries, however splendid and extraordinary, are not very likely to prove of any great benefit to the world, as they are rather objects of curiosity than of use.
Mrs. B: Such may be the illiberal conclusions of the ignorant and narrow-minded; but those who can duly estimate the advantages of enlarging the sphere of science, must be convinced that acquisition of every new fact, how-

ever unconnected it may at first appear with practical utility, must ultimately prove beneficial to mankind (*Fifth Edition* 355).

Caroline's dismissal sets up a tension that allows Mrs. B. to develop her evaluation. Mrs. B. then leads the girls from these larger issues to an account of Davy's demonstration. This account parallels Davy's closely, but interjects explanations of the use of platinum disks and naphtha as a storage medium, in response to Emily's questions and comments.

Mrs. B: The body which he first submitted to the Voltaic battery, and which had never yet been decomposed, was one of the fixed alkalis, called potash. This substance gave out an elastic fluid at the positive wire, which was ascertained to be oxygen, and at the negative wire, small globules of a very high metallic lustre, very similar in appearance to mercury; thus proving that potash, which had hitherto been considered as a simple incombustible body, was in fact a metallic oxyd; and that its incombustibility proceeded from its being already combined with oxygen.
Emily: I suppose the wires used in this experiment were of platina, as they were when you decomposed water; for if of iron, the oxygen would have combined with the wire, instead of appearing in the form of gas.
Mrs. B: Certainly: the metal, however, would equally have been disengaged. Sir H. Davy has distinguished this new substance by the name of POTASSIUM, which is derived from that of the alkali from which it is procured. I have some small pieces of it in this vial, but you have already seen it, as it is the metal which we burnt in contact with sulphur.
Emily: What is the liquid in which you keep it?
Mrs. B: It is naphtha, a bituminous liquid, with which I shall hereafter make you acquainted. It is almost the only fluid in which potassium can be preserved, as it contains no oxygen, and this metal has so powerful an attraction for oxygen, that it will not only absorb it from the air, but likewise from water, or any body whatever that contains it.
Emily: This, then, is one of the bodies that oxydates spontaneously without the application of heat?
Mrs. B: Yes, and it has this remarkable peculiarity that it attracts oxygen much more rapidly from water than from air; so that when it is thrown into water, however cold, it actually bursts into flame. I shall now throw a small piece, about the size of a pin's head, on this drop of water.
Caroline: It instantly exploded, producing a little flash of light! this is, indeed, a most curious substance!
Mrs. B: By its combustion it is reconverted into potash; and as potash is now decidedly a compound body, I shall not enter any of its properties until we have completed our review of the simple bodies (*Fifth Edition* 356–58).

Davy's text must be adapted to fit into the dialogue form. The turns are tightly linked by cohesive devices; in this selection, every turn is linked to the previous one by a pronoun (*it, this*), a demonstrative adjective (*this, these*), a comparative (*equally*), or a substitute form (*such*). The effect is that no turn makes sense by itself; one must keep looking back. The structure of the turn taking is, as one would expect, dominated by questions and answers. Again, the answers make sense only in relation to the preceding questions ("What is the liquid in which you keep it?"). As in the other dialogues, Emily and Caroline remain in character, the good student Emily asking the right questions at the right time while the troublesome Caroline responds to the show. While Davy's reader had to supply a reason for the selection of platinum, Emily gives it here: "for if [the wire were] of iron, the oxygen would have combined with the wire." When Emily notes that potassium "is one of the bodies that oxydates spontaneously," she points out the inference to be drawn from the set of facts they have just been given; this is the sort of move Brougham marked with logical connectors. The emphasis on oxygen throughout shows her pedagogical concern to repeat the main theme of her course; neither Brougham nor Davy need go back to the basics of the chemistry of the time.

As in Davy's and Brougham's texts, there is a shift here from exposition to experiment. But here there is also a shift from the real experiments of Davy to the fictional demonstration by Mrs. B., when she displays a piece of potassium. Note how the naming here goes with the demonstration, at the end of the description of Davy's work and before her own object lesson. As I have suggested, she is constrained in her demonstration, not by what could be described to this audience, but by the activities that would have been plausible at a country house. Thus she does not point to *this battery,* a large and expensive piece of apparatus available only to well-endowed laboratories like Davy's; she holds out *this vial.* But within this constraint, she dramatizes the effects. We see the effect through Caroline's exclamation, as she is converted from skepticism about Davy's work to wonder: "This is, indeed, a most curious substance!" Then Mrs. B. puts a damper on the enthusiasm by insisting that they stick to the announced plan of chapters.

Through the fiction, Marcet defines a forum in which she can take on chemistry while still remaining on terrain granted to women of her class. In *Conversations* she used a fictional form that enabled her to take on new areas of knowledge while protesting that she was not leaving home. Marcet's fiction both claims a domestic forum for women in chemistry, and pointedly accepts their exclusion from another, public forum. Her books strenuously deny that they are part of the public culture of science that the *Edinburgh Review* takes for granted, but through their fiction they move the subject to a different terrain. That terrain, the learning and conception of science in everyday life, is still being explored.

The crossing of these boundaries in fiction and demonstration is not limited to some nineteenth-century genres, nor is it a postmodernist conceit. It is part of the way we understand and divide the world. The making of a fiction is the mapping of one forum onto another, such as mapping the Royal Institution onto a country house. So it has social effects, just as the making of facts does. But the result is not just a blurring of the existing social divisions, or a utopian escape from the social assumptions that support them (men study and evaluate chemistry, women explain it to children). These fictions can constitute new textual forums with new relations within them. They explore what questions are asked, how they are answered, how and why people learn, what they relate the science to in their lives and enthusiasms.

One striking feature of all contemporary comments on Marcet, even those of other popularizers such as her friends Harriet Martineau and Mary Somerville, is their overwhelming condescension. Martineau says Marcet knew her place and kept to it.

> Mrs. Marcet never made any false pretensions. She never overrated her own books, nor, consciously, her own knowledge. She sought information from learned persons, believed she understood what she was told, and generally did so; wrote down in a clear, cheerful, serviceable style what she had to tell; submitted it to criticism gaily, and always protested against being ranked with authors of original quality, whether discoverers in science or thinkers in literature. (388–89)

Marcet's place was defined not only by the admitted limitations of her knowledge, but by her choice of audience, observational methods, and fictional form.

Just as we need to be skeptical about Martineau's estimate of Marcet's achievement, we need to be skeptical about her rather patronizing insistence on Marcet's modesty. Martineau is concerned to maintain the boundary between science and popularization, between authors of original quality and faithful conduits to a wider public. It is just this boundary that we must question. The construction of a popular forum for science was and is an important task. Perhaps our mistake has been to take Martineau's condescension and Marcet's own self-deprecation too seriously, or to listen to Mrs. B.'s scoldings, or Emily's humble questions, putting the learner in her place. Maybe to see the importance of this new forum, we need to see it more as did Caroline the troublesome pupil, impatient, resistant, and astonished.

Jane Marcet (1769–1858): Major Works

Conversations on Chemistry, Intended More Especially for the Female Sex (1806)
Conversations on Political Economy (1816)
Conversations on Natural Philosophy (1819)
Conversations on Vegetable Physiology (1829)
[Each of these books appeared in many British editions, and in many more pirate editions in the United States, and all were soon translated into French. Her later works for small children, such as *Conversations for Children on Land and Water* (1838), were popular in their time but are now very rarely found in academic libraries; they are listed in her entry in the *Dictionary of National Biography*, and there are copies of most of them in the British Library.]

Works Cited

Bazerman, Charles. *Shaping Written Knowledge: The Genre and Activity of the Experimental Article in Science*. Madison: U of Wisconsin P, 1988.

Benjamin, Marina. "Elbow Room: Women Writers on Science, 1790–1840." In *Science and Sensibility: Gender and Scientific Enquiry, 1780–1945*, ed. Marina Benjamin. Oxford: Basil Blackwell, 1991.

Brougham, Henry. Review of "The Bakerian Lecture on Some New Phenomena of Chemical Changes Produced by Electricity . . . ," by Humphry Davy. *Edinburgh Review*, 12 July 1808, 394–401.

Davy, Humphry. "Davy's Bakerian Lecture, on the Decomposition of Fixed Alkalies, &c." Rpt. in Jane Marcet, *Conversations on Chemistry*. New Haven, CT: Increase Cooke, 1813. 338–51.

Dear, Peter, ed. *The Literary Structure of Scientific Argument: Historical Studies*, Philadelphia: U of Pennsylvania P, 1991.

Foote, George. "Sir Humphry Davy and His Audience at the Royal Institution." *Isis* 43 (1952): 6–12.

Golinski, Jan. *Science as Public Culture: Chemistry and Enlightenment in Britain, 1760–1820*. Cambridge: Cambridge UP, 1992.

Henderson, Willie. "Harriet Martineau or 'When Political Economy Was Popular.' " *History of Education* 21 (1992): 383–403.

Knight, David. "Accomplishment or Dogma: Chemistry in the Introductory Works of Jane Marcet and Samuel Parkes." *Ambix* 33 (1986): 94–98.

Knight, David. *The Age of Science*. Oxford: Blackwell, 1986.

Lindee, M. Susan. "The American Career of Jane Marcet's *Conversations on Chemistry*." *Isis* 82 (1991): 8–23.

Marcet, Jane. *Conversations on Chemistry, In Which the Elements of That Science Are Familiarly Explained and Illustrated*. New Haven, CT: Increase Cooke, 1813.

Marcet, Jane. *Conversations on Chemistry, the Fifth Edition, Revised, Corrected, and Considerably Enlarged*. London: Longman, 1817.

Martineau, Harriet. *Biographical Sketches, 1852–1868*. 2d ed. London: Macmillan, 1869.

Myers, Greg. "Fictions for Exposition," *History of Science* 30 (1992): 221–47.
Myers, Greg. "Science for Women and Children: The Dialogue of Popular Science in the Nineteenth Century." In *Nature Transfigured: Science and Literature, 1700–1900*, ed. Sally Shuttleworth and J. R. R. Christie. Manchester: Manchester UP, 1989. 171–200.
Pavel, Thomas. *Fictional Worlds*. Cambridge: Harvard UP, 1986.
Ryan, Marie-Laure. "Possible Worlds and Accessibility Relations: A Semantic Typology of Fiction." *Poetics Today* 12 (1991): 553–76.
Selzer, Jack, ed. *Analyzing Scientific Prose*. Madison: U of Wisconsin P, 1993.
Shapin, Steven. "Pump and Circumstance: Robert Boyle's Literary Technology." *Social Studies of Science* 14 (1984): 481–520.
Shteir, Ann. "Botanical Dialogues: Maria Jacson and Women's Popular Science Writing in England." *Eighteenth Century Studies* 23 (1990): 301–17.
Shteir, Ann. "Botany in the Breakfast Room: Women and Early Nineteenth-Century British Plant Study." In *Uneasy Careers and Intimate Lives: Women in Science, 1789–1979*, ed. Pnina Abir-Am and Dorinda Outram. New Brunswick, NJ: Rutgers UP, 1987. 31–43.
Young, Robert. *Darwin's Metaphor: Nature's Place in Victorian Culture*. Cambridge: Cambridge UP, 1985.

4

Constructing Victorian Heavens
Agnes Clerke and the "New Astronomy"

Bernard Lightman

For Margaret Huggins, co-worker and wife of the eminent late Victorian astronomer Sir William Huggins, astronomy was "essentially a popular science." Of all the sciences, astronomy "appeals most readily and powerfully to the imagination" ("*System*" 383). But she perceived that increasing specialization, as well as the explosion of knowledge, required a new breed of popularizer who could interpret for the reading public the larger significance of late-nineteenth-century astronomical discoveries. "The progress of science and the growth of its literature during the last quarter of a century has been so enormous," Margaret Huggins declared, "that a new order of worker is imperatively called for" (*Agnes Mary Clerke* 15). The mission of this special worker is "to collect, collate, correlate, and digest the mass of observations and papers—to chronicle, in short, on one hand; and on the other, to discuss and suggest, and to expound" (*Agnes Mary Clerke* 17).

Strikingly, Huggins suggested that such a writer was now needed "to prepare material for experts," as well as to "inform and interest the general public" (*Agnes Mary Clerke* 17). The practicing scientist had no time to formulate new generalizations from the mass of accumulated facts and so was forced to pay heed "to a reviewing and digesting of things by those who are in his world indeed, but not of it; who have special knowledge and sympathy, but also leisure." What Huggins had

Figure 4.1. Agnes Mary Clerke, popularizer of the "New Astronomy." Source: M. L. Huggins, *Agnes Mary Clerke and Ellen Mary Clerke: An Appreciation* (printed for private circulation, 1907).

in mind here was so radically new that even she found it difficult to find "a proper name" for a class of workers in science who "must be historians, critics, something of experts, [and] something of originators all in one" ("*System*" 382). The conventional meaning of "popular writer" did not convey what she envisioned. Though she groped unsuccessfully for a new term, Huggins encountered no difficulty in coming up with the name of an individual who embodied all of the characteristics she associated with this new type of popularizer: her friend Agnes Mary Clerke (1842–1907), a prolific author of books, periodical essays, and encyclopedia articles on topics astronomical, ranging from the history of astronomy to current developments in astrophysics and stellar astronomy. Huggins was especially impressed by Clerke's ability to "stand midway between the lofty levels of highest attainment and speculation and the lower levels of average human capacity, interpreting the one to the other, and drawing both into closer union with a common All Father" ("*System*" 383). Unlike other writers who were reluctant to mention the word "God" where science was concerned, Clerke, a devout Catholic, was praised by Huggins for her determination to "show again and again in her book her deep reverence for the Deity—her faith in a divine order in which and toward which all things move" ("*System*" 386).

A comparison between Clerke and her "sisters in science" of the previous hundred years confirms Margaret Huggins's observation that she had redefined the role of female popularizer of science. Margaret Bryan (1797–1815), author of *A Compendious System of Astronomy* (1797) and other works on scientific subjects, ran a small school for girls at various locations in Blackheath and London. Unlike Clerke, Bryan worked in virtual isolation from the scientific establishment. Whereas Bryan presented informal student lectures in her texts (like other female popularizers of the period who used familiar forms of exposition such as letters and conversations), Clerke adopted the conventional format of the male scientist—the descriptive, impersonal, and objective scientific report. Bryan's chosen audience, the general public, was more in keeping with the aim of traditional female popularizers to instruct children and young women in the rudiments of scientific theory (Brody; Todd). But Clerke addressed adults rather than children, and male adults rather than young women. Clerke has more in common with Mary Somerville (1780–1872). Both were on familiar terms with the leading scientific men of England, and both produced best-sellers that appealed to a popular audience as well as scientists (Patterson x). But whereas Somerville set out to educate women, and in her *On the Connexion of the Physical Sciences* (1834) educated scientific specialists in one field to the current discoveries in another, Clerke aimed to educate astronomers in their own specialty. As an insider in astronomy, and as another woman interested in the popularization of science, Margaret Huggins had insight into the value of Clerke's contributions to astronomy and likely even into the perils of going beyond the prescribed role of the popularizer who reproduced the results of experts (Brück, "Companions" 70). Clerke, whether she wished to or not, posed questions about the role of women in science. Her anomalous position within the English astronomical world illuminates the complex range of choices that any woman faced when attempting to gain entrance into the male-dominated scientific community.

For British women interested in pursuing a career in astronomy, choices were limited (Kidwell 534–46). Virtually excluded from the observatory, college, and university, Clerke, Huggins, and other women interested in a focus for their astronomical activity were forced to look to scientific societies. Though the meetings of Section A (Mathematics and Physics) of the British Association for the Advancement of Science were open to women after 1838, participation in Britain's main professional forum for astronomy, the Royal Astronomical Society, was blocked for most women up until 1915, when qualified women were finally granted the privilege of membership (Dreyer and Turner 234). Women were admitted as honorary fellows from 1835, but in the nineteenth century only three received the honor—Caroline Herschel, Mary Somerville, and Anne Sheepshanks. However, Clerke's achievements could not be ignored, and in 1903, along with her friend Margaret Huggins, she was elected by the RAS Council as an honorary member of the society.

Born in western Ireland in 1842 into a well-to-do Anglo-Irish family, Agnes Clerke came into the world during the troubled times of the Irish famine.[1] Yet by all accounts the Clerke household was a happy one. Clerke's older sister Ellen Mary (born 1840) was her lifelong companion and shared her fascination with astronomy. The Clerke children's love of science was no doubt fostered by their father, a bank manager, who provided "an environment of scientific suggestion" during their early lives (Huggins, *Agnes Mary Clerke* viii). Science was a family activity. A brother, Aubrey, recalled with amusement the wrath of the family's Irish servant when the evil-smelling odors from his father's chemistry experiments escaped the confines of the laboratory. He also remembered the glimpses of Saturn's rings and Jupiter's moons through his father's telescope mounted in the garden (Huggins, *Agnes Mary Clerke* vii). Clerke's early education was provided entirely by her scholarly parents, who instilled within her the joy of learning while giving her the ability to conduct private research. By the age of eleven she had mastered Herschel's *Outlines of Astronomy,* which only whetted her appetite to study astronomy in more depth (Brück, "Ellen and Agnes Clerke" 29). Four years later she was already thinking of writing a history of astronomy, and may even have composed a few chapters (Huggins, *Agnes Mary Clerke* 9).

Later, in 1877, after moving to Dublin and then to Italy, the entire Clerke family settled in London, where Agnes began to pursue a career as a writer. Agnes Clerke's output is daunting—it also singles her out as one of the great popularizers of science in the late Victorian period. The success of *A Popular History of Astronomy during the Nineteenth Century* (1885) first brought her to the attention of the scientific community. Her other major works included *The System of the Stars* (1890), *The Herschels and Modern Astronomy* (1895), *Astronomy* (1898), of which she was co-author, *Problems in Astrophysics* (1903), and *Modern Cosmogonies* (1905). With her prolific pen Agnes Clerke also penned scores of essays, mostly on astronomy, for periodicals like the *Edinburgh Review, Knowledge, Observatory,* and *Nature*. For *The Dictionary of National Biography* she produced 149 biographies of famous astronomers, and she submitted entries on astronomers and astronomical subjects to *The Encyclopedia Britannica*. As a result of her labors, Clerke gained partial admission into the male-dominated astronomical world. Among her correspondents were the Hugginses, Norman Lockyer, David Gill, E. S. Holden of the Lick Observatory in California, and E. C. Pickering of Harvard (Kidwell 543). One American astrophysicist considered Clerke to be such an important molder of scientific opinion that he often informed her first of his discoveries (Osterbrock 315).

Both her contemporaries and twentieth-century scholars have regarded Clerke's *Popular History of Astronomy during the Nineteenth Century* as an authoritative secondary source on the history of nineteenth-century science and tend to emphasize Clerke's role as historian.[2] Indeed, the rigorous scholarly training that Clerke received from her parents is reflected in her determination to stick to "the original authorities" and "take as little as possible at second-hand" (Clerke, *Popular* vii). Using primary sources Clerke set out in her *Popular History* to tell the story of the "new astronomy" that came into existence during the nineteenth century.

According to Clerke, scientists had developed three kinds of astronomy, "each with a different origin and history, but all mutually dependent and composing in their fundamental unity, one science" (1). The first and oldest branch, created by astronomers from ancient times up to Copernicus, Clerke labeled "observational" or "practical," since it was primarily concerned with noting facts as accurately as possible without inquiry into their connection or cause. Newton founded the second kind, "gravitational" or "theoretical" astronomy, which focused on a single law as cause of the intricate relationships between the heavenly bodies. What distinguished the third division of celestial science, termed by Clerke "physical and descriptive astronomy," was its emphasis on knowing the nature or constitution of the heavenly bodies. Inquiries of this kind were made possible by the use of the spectroscope and photographic camera (1–2). The "new astronomy" had not only revitalized celestial science but also served as the moving force behind the unification of the physical sciences. Through the use of their new scientific instruments astronomers could apply physics and chemistry to the study of the heavens and in turn open up a new understanding of the physics and chemistry of terrestrial processes (9). Clerke was enthusiastic about the potential of this "new astronomy." "It promises everything," she declared; "it has already performed much; it will doubtless perform much more" (183).

In her own works, Clerke positioned herself more often as a popularizer than as a historian. In the preface to the first edition of *The System of the Stars,* Clerke asserted that her object was to bring the results of sidereal science "within the reach of many," argued that "astronomy is essentially a popular science" to which "the general public has an indefeasible right," and insisted that astronomy be kept clear of the baleful consequences of specialization, "unnecessary technical impediments," through "literary treatment" (ix). Fortunately, the "new astronomy" had rendered "the science of the heavenly bodies more popular" in its "nature." The kind of knowledge accumulated by the "new astronomy," gathered through spectrum analysis and camera, was less remote from ordinary experience, which facilitated the task of describing recent astronomical discoveries in simple, intelligible language (*Popular* v–vi).

In addition to explaining to the reading public the operation of, and data collected by, the new scientific instruments, Clerke also stressed the larger cosmic significance of the "new astronomy." One of the major goals of Clerke's popularization of the "new astronomy" was to demonstrate that it is susceptible of a powerful religious interpretation, a task reflecting her concerns as an English Catholic living in a society in the throes of a vast process of secularization. Clerke tried to prevent scientists hostile to Christianity from using scientific theories to undermine the power of religion. Likening Haeckel's evolutionary monism to the Renaissance animism of Campanella, she contemptuously leaves such views "to refute themselves" ("Campanella" 162). Clerke also tried to counter the pernicious effects of new scientific ideas, like entropy, which seemed to suggest that humans were trapped in a meaningless, material universe. Clerke told those distraught over the depressing predictions of physicists that the universe was slowly winding down that she rejected pessimism. Whatever destiny God intended for the human race, astronomers had determined that the solar system would last long enough for its accomplishment (*Modern* 245). Clerke reminded her readers that "the extreme conclusions of science are invariably pessimistic, because they are reached without taking any account of the intelligent control perpetually, though insensibly, overruling the workings of blind forces" (Clerke, Fowler, and Gore 233). The physicist's idea of universal heat death held no terror for one who recognized that "He who made the heavens can restore them" (*System* 375). Clerke's effort to put the "new astronomy" into a religious framework should be seen as part of her larger project to "renovate intellectual life" so that thinking individuals did not feel compelled to choose between science and religion ("Campanella" 168).

The link between astronomy and religious faith was made explicit by Clerke throughout her works. In the preface to the first edition of *A Popular History,* she identified as one of her goals the attempt to help readers "towards a fuller understanding of the manifold works which have in all ages irresistibly spoken to man of the glory of God" (vi). Wherever she turned Clerke saw God throughout the entire gamut of astronomical phenomena. Whether it be the evolution of the planets, whose growth is guided "from the beginning by Omnipotent Wisdom"; or the "sequence of Divinely decreed changes" by which nebulae are transformed into star clusters; or even gigantic galactic rifts of starless space, wherein "Supreme Power is at work in dispersing or refashioning" star clouds, Clerke sees the hand of God (*Popular* 348; *System* 207; *Problems* 541). At least since the time of Newton, astronomy provided the most powerful scientific aid to faith, for no matter where the telescope was pointed it revealed the same pattern of design in the limitless regions of space that was so evident on earth (*Popular* 19). The face of the skies may change, not in a haphazard fashion, "but in subjection to laws unalterable in their essence although infinitely various in their applications, divinely directed towards the continually more perfect embodiment of the unfolding Eternal Thought" (*System* 375). Clerke was as awestruck by the design and order in nature as any pre-Darwinian natural theologian.

However, Clerke recognized that natural theology could not continue unaltered, especially after the revelations of the "new astronomy." She aimed to present an updated version informed by the most recent astronomical discoveries. At the end of the eighteenth century, astronomers conceived of the universe as a majestic and symmetrical machine characterized by simplicity, harmony, and order. But in addition to widening our knowledge of the heavens, the spectroscope and the camera had revealed a divinely designed universe that was characterized by complexity and inexhaustible variety (*Popular* 452). "The heavens are full of surprises," Clerke exclaimed, after puzzling over changes of brightness in variable nebulae (*Problems* 522). The uncertain origin of certain types of stars leads Clerke to conclude that "nature does not run in a groove" (*Problems* 279). At times, Clerke could point to discoveries of the "new astronomy" that disclosed orderly patterns in the heavens hitherto concealed by apparent chaos. The "seeming confusion" of asteroids when compared with the "harmoniously ordered and rhythmically separated orbits of the larger planets" proved to be "not without a plan" (*Popular* 285). The distribution of stars and nebulae is "easily seen to be the outcome of design" (*Modern* 147). The "new astronomy," which was slowly bringing about a unification of science, revealed and reflected the unity of design behind nature. In breaking down the barriers between physics and astronomy, terrestrial and celestial science, the earth and the stars, the "new astronomy" was a science that "aims at being one and universal, even as Nature—the visible reflection of the invisible highest Unity—is one and universal" (*Popular* 142).

But at other times Clerke admitted that even the "new astronomy" threw no light on the order lying behind particular celestial phenomena. After examining the mysterious clustering power of stars, Clerke declared that "a creative purpose can be *felt*, although it cannot be distinctly followed by the mind" (*Herschels* 175). Clerke often ends her chapters with a profession of ignorance and an affirmation that God lies behind impenetrable mystery. For example, after relaying to the reader all of the new discoveries concerning comets while pushing the evidence as far as it will go toward uncovering order, Clerke closed by offering tentative speculations as to the origins of comets. "They are, perhaps," she asserted, "survivals of an earlier, and by us scarcely and dimly conceivable state of things, when the swirling chaos from which the sun and planets were, by a supreme edict, to emerge, had not as yet separately begun to be" (*Popular* 371). Clerke concluded the chapter on stars and nebulae by allowing that the laws of the association of stars may be discovered one day, but "the economy of the higher order of association . . . will possibly continue to stimulate and baffle human curiosity to the end of time" (*Popular* 427). The book *Modern Cosmogonies* was brought to completion with a powerful argument for the limits of scientific knowledge and the insinuation conveyed to us by our consciousness of "a Power outside nature continually and inscrutably acting for order, elevation, and vivification" (282). Clerke's faith in divine design and natural law was so great that she affirmed its existence even where the "new astronomy" could not be summoned to ferret it out.

Clerke's writing style was in keeping with her efforts to communicate the cosmic significance of the "new astronomy" to her popular audience. She retained the impersonal, objective point of view of the typical scientific text, thereby preserving the authority of her work. But rather than losing the reader in dry recitation of facts and data, she emphasized the larger meaning of scientific discoveries. She often preferred to engage her audience's aesthetic sensibilities, drawing on colorful metaphors to illustrate a scientific point and eschewing overly technical language. For example, she explained why astronomers were so obsessed with designing bigger and better telescopes by referring to light as "the chief object of astrophysicist's greed" (*Popular* 432). In a passage from the last chapter of *Problems in Astrophysics* dealing with gigantic galactic clefts (large areas in space where there are no stars), Clerke created a dramatic mood, even though her subject is, almost literally, "nothing":

> [The galactic clefts are] intervals of starless space between neighbouring star-clouds, and suggest processes of disintegration advancing with inconceivable slowness towards unimagined issues. These wonderful collections are then in a state of flux; they are passing from one condition to another; Supreme Power is at work in dispersing or refashioning them, sending abroad their aggregated suns like flying sparks from the anvil. (541)

Structured to bring out the meaning in "nothing," this passage creates a mood of awe and reverence for nature as God's creation through the use of vivid metaphors and religious language.

According to Margaret Huggins, Clerke's role as popularizer involved more than lucid explanations of the theories and instruments of the "new astronomy," more than outlining the history of its successes, and even more than spelling out its dramatic religious implications. As skillful as Clerke may have been at collating, interpreting, and summarizing astronomical data, Huggins also saw her as a critic, an originator, and an expert. As critic, Clerke did not hesitate to dismiss a theory if she found it wanting or to criticize an eminent scientist if she disagreed with him. She rejected Lockyer's theory on the world-building function of meteorites as largely speculative, while Kelvin's notion of an extraterrestrial origin for earthly life was deemed less credible than the adventures of Baron Munchausen (*Modern* 126, 281). Besides her insights into the religious significance of the "new astronomy," Clerke performed the role of originator through her ability to make novel connections between the discoveries of isolated specialists. Clerke remarked that "a larger synthesis is demanded for the harmonizing of multitudinous facts, at present grouped incongruously, or left in baffling isolation; and it is rendered increasingly difficult of attainment by the continual growth of specialization" (*Modern* 160). Clerke took the results of scientists, synthesized them into a system, and gave the discipline of astronomy a new shape.

Huggins's claim that Clerke was something of an expert was substantiated by her contemporaries. Clerke was not merely a mediator between the scientific experts and the uninformed public but also stood as an interpreter of the larger meaning of recent astronomical discoveries to the professional astronomers themselves. After publishing *A Popular History,* Clerke began to work on projects that were less accessible to a popular audience than historical studies. One of the reasons that Clerke accepted David Gill's invitation to visit him at the Cape of Good Hope observatory in 1888 was to experience practical astronomical work firsthand under the guidance of a respected authority. Gill advised her that practical work would raise the level of expertise in her future books and preclude mistakes noticeable to a professional scientist (Forbes 199–200). Both *The System of the Stars* and *Problems in Astrophysics* examined cutting-edge issues in astronomy, emphasizing the present state of knowledge and future paths of investigation, rather than recount the past discoveries of eminent scientists. A typical chapter in *Problems in Astrophysics,* which dealt in great technical detail with the sun, stars, and nebulae, reviewed all of the most recent research in the area, acknowledged the many questions remaining, and suggested work to be done by astronomers to begin answering these questions. Clerke instructed experts which lines of research to pursue. Clerke's correspondent and American friend E. S. Holden clearly perceived her intentions in *The System of the Stars.* Congratulating Clerke on the publication of the book in 1891, he observed that "it is not only an adequate account of our present knowledge but it *is* full of pregnant suggestion for our future guidance" (E. S. Holden to Agnes Clerke, 21 Jan. 1891, Shane Archives). Always encouraging throughout their correspondence, two years earlier Holden had voluntarily written to the president of Vassar College to nominate Clerke as successor to the late Maria Mitchell as director of the observatory, asserting that "there are very few living men who have her [Clerke's] philosophical grasp of the whole science and very few of equal erudition" (E. S. Holden to the President and Trustees of Vassar College, 6 Aug. 1889, Shane Archives).

Clerke's anomalous position within the astronomical community can be detected in contemporary reviews of her books. Some reviewers, like Margaret Huggins, were willing to concede that Clerke represented a novel development within astronomy. In the *Nation* an anonymous reviewer affirmed that Clerke's *Problems in Astrophysics* was "not a book of popular science: it is a popular book on professional science—a thing seldom to be found" ("Clerke's Astrophysics" 98). Two important astronomers echoed this view, E. W. Maunder, who said that Clerke's *Popular History* would be valued by the astronomer, not just the general reader, and Robert S. Ball, who believed that the same book should not be classed as a "popular work in the ordinary acceptation" (Maunder 126; Ball 313). However the majority of the reviewers, though favorably disposed toward her books, tended to deny Clerke's expertise or critical skills, and relegated her to the roles of historian or traditional popularizer. She was deemed "too sparing of criticism" in *Knowledge* and damned by W. H. Wesley, longtime assistant secretary of the Royal Astronomical Society, for carrying "the virtue of impartiality to excess" ("Notices of Books" 33; Wesley 36). Clerke's expertise as scientist was questioned in the pages of the *Nation*, where the reviewer complained of the regrettable lack of tables and algebraical formulae; by Ball, who pointed to remarks in Clerke's *Popular History* that could not have been made by a practical astronomer; and by Maunder, who asserted that Clerke did not seem to "have had an observer's training" ("Clerke's Astrophysics" 99; Ball 313; Maunder 128). "Miss Clerke is essentially an historian of astronomy," a reviewer in *Knowledge* declared, sounding a familiar theme in numerous reviews ("Notices of Books" 33). Like many others, R. J. Mann, a popularizer in his own right who wrote a long series of popular textbooks on astronomy, chemistry, and physiology, conferred on Clerke "the mantle of Mrs. Somerville," which placed her in the category of traditional popularizer (372).

Widespread commentary on Clerke's writing skills, though usually positive, implicitly relegated her to the role of nonexpert popularizer and became a means for some reviewers to connect literary style to gender. Ball and Maunder praised Clerke's literary style, while in the pages of *Knowledge* readers were told that no writer of Clerke's time "had a juster sense of style" (Ball 314; Maunder 128; E. S. G. 32). In his comparative review of *A Popular History* and Ball's *The Story of the Heavens*, Mann credited Ball with a stronger grasp of astronomical science, but opined that his book lacked "the finish and grace with which the less masculine 'Popular History' has been worked out" (402). Reviewers' reactions to Clerke's literary prowess underline a basic paradox in her popularization project: by using "feminine" literary devices to draw in a popular audience, Clerke lost the right to be taken seriously as an expert who could address scientists on their own terms.

In reviews of Clerke's more technical works, Richard A. Gregory (1864–1952), Norman Lockyer's protégé and assistant editor of *Nature,* launched into a vicious attack on Clerke's scientific credentials that explicitly raised the issue of gender. Gregory maintained in his review of *Problems in Astrophysics* in 1903 that Clerke did not understand the "real nature of some of the material collected" and that "she passes judgment and gives advice on matters which can only be rightly understood by investigators actively engaged in spectroscopic work" ("Spectroscope" 339). Like reviewers favorably disposed toward Clerke, Gregory refused to grant Clerke the status of expert popularizer who spoke to scientists as equals and placed her in the role of historian. But whereas other reviewers were impressed with Clerke's abilities as historian, Gregory thought she failed here as well. Clerke went beyond the proper role of the historian who assimilated and described, "irritating the men who are doing the work by expressing her opinion upon it or suggesting what course they ought to take next" ("Spectroscope" 339). Gregory's hostility is due, in part, to his belief that Clerke had not given enough credit to the contributions of his patron, Lockyer, to astrophysics ("Spectroscope" 340–41). But he also took offense that a woman dared to play the impartial scientific critic, a task precluded by her nature:

> In preparing a statement of the position of fact and theory in any branch of science, great care must be exercised, and not a single assertion should be made without substantial reason for it. A cynic has said that it is a characteristic of women to make rash assertions, and in the absence of contradiction to accept them as true. Miss Clerke is apparently not free from this weakness of her sex. ("Spectroscope" 339)

Gregory effectively excluded Clerke, and virtually all women, from engaging in scientific work.

Three years later Gregory reviewed a new edition of *The System of the Stars* and again based his objections to her work on gender considerations. Acknowledging that it was regarded as "ungracious" in "these days" to make the suggestion that the "intuitive instinct of a woman is a safer guide to follow than her reasoning faculties," Gregory nevertheless declared that "evidence of its truth is not difficult to discover in most literary products of the feminine mind." It was no disparagement, Gregory claimed, to say that even Clerke "shares this characteristic of her sex," for a woman's intuition is an attractive instinct when applied to ordinary affairs of life. However, "when it influences the historiographic consideration of contributions to natural knowledge," it is a fatal weakness. Gregory reminded his readers that Clerke was "not actively engaged in the investigation of the field surveyed" and that she was a "bibliographer" whose work should be judged accordingly ("Stars" 505). He judged it to be inadequate. He portrayed Clerke as a promiscuous female who "flirts" with exotic scientific concepts "whenever she has the opportunity though little is known of their resources" (507). Despite Gregory's sarcastic reference to those "who follow the trend of a writer like Miss Clerke with lamb-like sequacity" (505), open hostility to Clerke was rare. However *Nature* was an influential and widely read journal.

The phrase "the construction of the heavens," which Clerke found in William Herschel, aptly encapsulates her role as popularizer (*Herschels* 53). For Clerke, as for Herschel, the "construction of the heavens" was the work of a divine being. But Clerke engaged in the gigantic undertaking of constructing the heavens in opposition to the constructions of those who claimed that science in the modern age must be cleansed of its religious associations. Her task was to demonstrate to scientists and the reading public how the camera and the spectroscope had revealed new facets of a design in a universe of boundless variety. As Margaret Huggins had observed, this lofty goal required a new sort of popularizer who was historian, critic, expert, and originator all rolled into one. Ironically, Clerke, the new-style popularizer, perpetuated in her astronomical work the older, natural theology tradition that was so important in English thought previous to Darwin. But in redefining the role of popularizer, Clerke also ran afoul of male scientists like Gregory, who denied the validity of her construction of the heavens on the grounds that all women are, by their very nature, prevented from thinking impartially or rationally.

Agnes Clerke (1842–1907): Major Works

A Popular History of Astronomy during the Nineteenth Century (1885)
The System of the Stars (1890)
The Herschels and Modern Astronomy (1895)
Astronomy (1898) [co-authored with A. Fowler and J. Ellard Gore]
Problems in Astrophysics (1903)
Modern Cosmogonies (1905)

Notes

The author would like to express his gratitude to colleagues who provided useful information and helpful suggestions for this project, including Dr. Barbara Becker, Dr. M. T. Brück, Dr. Robert W. Smith, Dr. Peggy Kidwell, and Kenneth Weitzenhoffer. I also appreciated receiving permission from the Royal Astronomical Society and the Lick Observatory to publish material from their archives. A research trip to London was funded through a York University Faculty of Arts Research Grant. The work for this essay was done while the author held a Research Grant from the Social Science and Humanities Research Council of Canada.

1. For this biographical section I am heavily indebted to the work of Mary Brück, which complements and supplements Lady Huggins's biography.

2. Shortly after Clerke's death F. W. Levander, the president of the British Astronomical Association, affirmed that the book "had always been looked upon as a classic" ("Report of Meeting of the Association" 155). In her obituary notice Margaret Huggins referred to *The Popular History* as the "standard work" in the history of astronomy (230). More recently A. J. Meadows recommends Clerke in his select bibliography as a good source of information on astronomical studies of the sun (311). Naturally, the dearth of historical surveys of nineteenth-century astronomy has led scholars to rely on Clerke.

Works Cited

Ball, Robert S. "Astronomy during the Nineteenth Century." *Nature* 33 (4 Feb. 1886): 313–14.

Brody, Judit. "The Pen Is Mightier Than the Test Tube." *New Scientist* 105 (14 Feb. 1985): 56–58.

Brück, Mary T. "Agnes Mary Clerke, Chronicler of Astronomy." *Quarterly Journal of the Royal Astronomical Society* 35 (1994): 59–79.

Brück, Mary T. "Companions in Astronomy: Margaret Lindsay Huggins and Agnes Mary Clerke." *Irish Astronomical Journal* 20 (Sept. 1991): 70–77.

Brück, Mary T. "Ellen and Agnes Clerke of Skibbereen, Scholars and Writers." *Seanchas Chairbre* 3 (1993): 23–42.

[Clerke, Agnes M.] "Campanella and Modern Italian Thought." *Edinburgh Review* 149 (1879): 139–68.

Clerke, Agnes M. *The Herschels and Modern Astronomy.* London, Paris, and Melbourne: Cassell, 1895.

Clerke, Agnes M. *Modern Cosmogonies.* London: Adam and Charles Black, 1905.

Clerke, Agnes M. *A Popular History of Astronomy during the Nineteenth Century.* Edinburgh: Adam and Charles Black, 1885.

Clerke, Agnes M. *Problems in Astrophysics.* London: Adam and Charles Black, 1903.

Clerke, Agnes M. *The System of the Stars.* 2d ed. London: Adam and Charles Black, 1905.

Clerke, Agnes M., A. Fowler, and J. Ellard Gore. *Astronomy.* New York: D. Appleton, 1898.
"Clerke's Astrophysics." *Nation* 77 (30 July 1903): 98–99.
Dreyer, J. L. E., and H. H. Turner, eds. *History of the Royal Astronomical Society, 1820–1920.* London: Royal Astronomical Society, 1923.
E. S. G. "The Late Miss Agnes Clerke." *Knowledge* 4 (1907): 31–32.
Forbes, George. *David Gill: Man and Astronomer.* London: John Murray, 1916.
Gregory, R. A. "The Spectroscope in Astronomy." *Nature* 68 (13 Aug. 1903): 338–41.
Gregory, R. A. "Stars and Nebulae." *Nature* 73 (29 March 1906): 505–8.
Huggins, M. L. *Agnes Mary Clerke and Ellen Mary Clerke: An Appreciation.* Printed for private circulation, 1907.
Huggins, M. L. "Obituary—Agnes Mary Clerke." *Monthly Notices of the Royal Astronomical Society* 67 (Feb. 1907): 230–31.
Huggins, M. L. "*The System of the Stars.*" *Observatory* 13 (Dec. 1860): 382–86.
Kidwell, Peggy. "Women Astronomers in Britain, 1780–1930." *Isis* 75 (1984): 534–46.
[Mann, R. J.] "The Recent Progress of Astronomy." *Edinburgh Review* 163 (April 1886): 372–405.
Maunder, E. W. "History of Astronomy in the Nineteenth Century." *Observatory* 9 (March 1886): 126–28.
Meadows, A. J. *Science and Controversy: A Biography of Sir Norman Lockyer.* London: Macmillan, 1972.
"Notices of Books—*The System of the Stars.*" *Knowledge* 14 (2 Feb. 1891): 32–33.
Osterbrock, Donald E. *James E. Keeler: Pioneer American Astrophysicist and the Early Development of American Astrophysics.* Cambridge: Cambridge UP, 1984.
Patterson, Elizabeth Chambers. *Mary Somerville and the Cultivation of Science, 1815–1840.* Boston/The Hague/Dordrecht/Lancaster: Martinus Nijhoff, 1983.
"Report of Meeting of the Association Held on January 30, 1907, at Sion College, Victoria Embankment." *Journal of the British Astronomical Association* 17 (1906–7): 155–64.
Shane Archives. Mary Lea Shane Archives of the Lick Observatory, University of California, Santa Cruz.
Todd, Janet, ed. "Bryan, Margaret." *A Dictionary of British and American Women Writers, 1660–1800.* Totowa, NJ: Rowman and Allanheld, 1985. 62–63.
Wesley, W. H. "Notice of Book—*History of Astronomy during the Nineteenth Century.*" *Knowledge* 18 (1 Feb. 1895): 34–36.

PART 4

DISSEMINATING KNOWLEDGE: CANADA, AUSTRALIA, AND AMERICA

5

Science in Canada's Backwoods
Catharine Parr Traill

Marianne Gosztonyi Ainley

By the time thirty-year-old Catharine Parr Traill (née Strickland) sailed from Britain to Canada in 1832, the study of natural history and popular science writing had become accepted occupations for upper-class European women. She brought with her a keen interest in science and many years of experience as a writer for the British public. Her readers included children and adults (men and women), and throughout her long and productive life she continued to write for mixed audiences. While Catharine Parr Traill never had the extensive leisure time nineteenth-century men seemed to need for scientific activity, during the sixty-seven years she lived in Upper Canada she found many opportunities to study natural history and to integrate her knowledge of science into her writing. She was a prolific writer both by inclination and by necessity.

In the English-speaking world, as Cole, McCallum, and Needler have shown, Catharine Parr Traill was known to her contemporaries as a naturalist, botanist, and writer; in the second half of the nineteenth century, science and literature were still considered as part of general culture rather than mutually exclusive activities. In twentieth-century Canada her scientific writing has been minimized for several reasons: the lack of an appropriate context for an (en)gendered history of Canadian science or for women's science writing, and the androcentric per-

Figure 5.1. Catharine Parr Traill (1802–1899). Photograph courtesy of Dr. Elizabeth Parnis.

spective of well-known male scholars such as Ballstadt and Berger which prevented them from acknowledging her contributions to the supposedly male domain of science. Thus, she was considered, in Peterman's words, a "splendid anachronism," and "a struggling amateur," instead of a pioneering naturalist and popularizer of science.[1] Now, in the 1990s, it is time to reevaluate her work.

Catharine Parr Traill's Canadian writing career spanned more than six decades. Her prodigious output masks the fact that much of the time she wrote under extremely difficult circumstances. She had a privileged upbringing yet lived most of her life near poverty. She and her five sisters were educated by their parents while living on their country estates Stowe House and Reydon Hall, or in the town of Norwich (all of them in East Anglia). As a child she learned to observe nature, to collect, label, and identify botanical specimens, and she developed the ability to think critically. The Strickland sisters read widely, including books by popular women writers. Catharine Parr was also influenced by the natural history books of Izaak Walton and Gilbert White and the fiction of Daniel Defoe and Thomas Day. Her broad interest in science was evident in her early work, and throughout her long life she incorporated scientific observations and information into books written for emigrants and children. In addition she produced more specialized works on botany and natural history, written for both the general public and other naturalists. In England, during the 1820s, she became fascinated with the New World and included information about the natural history of eastern North America in her writing.

In 1831 she met Thomas Traill (1793–1859), a Scot, who was educated at Oxford and planned to emigrate to Canada. They married in May 1832 and, prior to their departure for Canada in July, spent their brief honeymoon in Scotland. On the way to the Orkneys, the couple met a botany professor from Edinburgh University (probably Robert Graham), who asked Catharine Parr Traill to collect plants for him in Canada.[2]

In 1832, what was the status of western science in what is now called Canada? What were the possibilities for a woman with Catharine Parr Traill's interests to study and write about science? We know that early scientific studies in Canada were carried out by French and British explorers and colonizers, army and navy personnel, civil servants, and, to a lesser extent, missionaries. Most eighteenth- and nineteenth-century natural history studies were based on collections and short-term observations. Western science was "colonial" because, as Basalla and others have shown, the center for scientific activity was in Europe and topics for investigation were defined by European scientists.[3] French, English, and Swedish publications on the natural history of eastern North America came from two major sources. One, as I have shown, was the environmental knowledge transmitted to Hudson's Bay Company personnel by native women and men. The other, as Chartrand, Farber, Levere and others have noted, was the collections, observations, and descriptions of European trained men—explorers, missionaries, and colonizers—and a few women like the missionary

Marie de l'Incarnation, the herbalist Catherine Jérémie, and the French collector Mme. de Bandeville.

By the early 1830s, science in Upper and Lower Canada was in a state of transition. There were few institutions of higher learning even for men. According to Croft, science writing as an occupation was still rare. In the 1820s, in Lower Canada, Lady Dalhousie (née Christian Broun, 1786–1839) and her friend Harriet Campbell Sheppard (d. 1877) had published scientific articles in the *Transactions* of the Literary and Historical Society of Quebec. Their friend, Anne Mary Perceval (née Flower, 1790–1876), contributed to botanical correspondence and collections, but did not publish. These three women were well-educated members of the British leisure class who studied plants, shells, birds, and native customs, and corresponded with and sent specimens to British and North American naturalists, such as W. J. Hooker, John Torrey, and William Darlington (Pringle 10; Ainley, "Last in the Field?"). Their science writing and their scientific collections were produced for other naturalists. They did not write for a wider audience, and they did not get paid for their writing. Other women also made scientific collections and attended lectures in Lower Canada (now Quebec). By contrast, there were hardly any practitioners of natural history in Upper Canada, now Ontario. As Zeller and Waiser have pointed out, the utilitarian aspects of science, such as agriculture, geology, mining, and metallurgy, were strongly supported by the new Scottish and United Empire Loyalist settlers. By contrast, natural history studies and explorations found little support (Ainley, "From Natural History"; Zeller 46–48). In Upper Canada, Catharine Parr Traill was to use her new discoveries as subject matter for her books, which were directed originally at receptive British readers, and later also at the Canadian and American public. Her topics were based on her own impressions and experiences as an upper-middle-class emigrant woman—wife and mother—in the backwoods. She was open-minded and willing to learn about the life of settlers, a life far different from anything she had experienced in England. The sights and smells of the changing landscape during her ocean voyage, inland boat trips, and uncomfortable rides on rough roads through the recently cleared regions of Upper Canada formed a part of the letters she sent home. Other vivid recollections reflected her encounters with the Canadian Shield, the varied environment of lakes, drumlins, eskers, and ancient rock outcrops left by the retreat of ice during the last ice age. In *The Backwoods of Canada* she later recalled her first impressions of "the bold forest of oaks, beech, maple and basswood, with now and then a grove of dark pine . . . the exquisite flowers and shrubs . . . many [of them] peculiar to the plains" (55), and the

Otonabee river "a fine, broad, clear stream . . . between thickly wooded banks" (63).

The Traills initially stayed in Peterborough with Catharine's brother Samuel who had settled in the district in 1825. While waiting to move to their own land, Catharine Parr Traill explored the district and collected flowers, lichens, and fungi. Soon in her first home as a married woman (a simple log house), she embarked on her unimaginably hard life as a settler, and, at the same time, her long and satisfying career as a naturalist and science writer. In spite of the grueling work involved in running a pioneer household, she kept a diary, collected flowers, studied the natural history of the region, and began to write with an eye to the paying public. For the next twenty years, her older sisters in England served as her editors and literary agents.

The Backwoods of Canada was based on Catharine Parr Traill's early Canadian experiences. It sent a positive message to prospective middle- and upper-class emigrants, but did not deny the difficulties faced by new settlers. The book was favorably received in England. By 1840, it had five printings and, in 1846, two new editions—all in England. Apparently, there were also a number of pirated editions both in England and in the United States. In the meantime, her Oxford-educated husband lost interest in farming. Like other educated gentlemen, Thomas Traill was singularly unable to deal with the hardships encountered by pioneer settlers. By contrast, Catharine Parr Traill, who in early life was pampered by her parents and siblings, was tough, resourceful, curious, adaptable, and remained cheerful. She had a strong, evangelical Anglican faith (evident even in her science writing) that, combined with her resourcefulness, helped her to overcome difficulties. As a young woman in England she obtained independence through her income from writing. In Canada, while she bore nine children to an often depressed husband, and endured worsening arthritis, "Mrs. Traill" continued to write for a living. By the late 1840s, when the Traills moved to Rice Lake, she became the chief breadwinner of the family. At Oaklands, Rice Lake, Catharine Parr Traill wrote two children's books, *Lady Mary and Her Nurse* (1850) and *Canadian Crusoes* (1852), and sold their copyrights for fifty pounds sterling to the British publisher Hall and Virtue. The books were reprinted in other editions, under various names, but according to FitzGibbon, Morris, and Eaton, she received no further income from them.

Catharine Parr Traill's familiarity with up-to-date works on British natural history was of limited use during her first few years in Upper Canada. Yet, like many of her contemporaries she remained interested in all aspects of natural history. In *The Young Emigrants* and *The Back-*

woods, she wrote about rocks, fossils, and minerals, in addition to plants and animals. There is no evidence, however, that in her later years Catharine Parr Traill read as much about geology as she did about zoology or botany. In the "backwoods," she encountered plants and animals unique to North America and others that were similar to their European relatives. From a new friend, Frances Stewart (née Browne, 1794–1872), Catharine Parr Traill borrowed Frederick Pursh's *Flora Americae Septentrionalis* (1814), which she deciphered with the help of her husband. She had no access to comparable works on North American animals and initially learned about them from her own observations and from the stories of natives and older settlers.

In Upper Canada, Catharine Parr Traill found a freedom of action and thought that enabled her to know her own position, to state her opinion, and challenge those of others. In contrast to her earlier *The Young Emigrants* which had a Canadian topic but was based on the experience of others, in *The Backwoods* she relied upon her own observations of plants and animals. In *The Young Emigrants,* the young author had uncritically repeated erroneous information, because she had no opportunity to make firsthand observations. She wrote there of the American [belted] kingfisher "whose splendid plumage far surpasses any of those seen in England" (49) and about the many species of hummingbird of a most brilliant colour" (50). The older, more critical Catharine Parr Traill would not have agreed with this description of the duller, bluish-grey North American kingfisher and knew that there was only one species of hummingbird in Upper Canada. In *The Backwoods,* she compared the size, song, and general habits (behavior) of various Canadian and English birds and frogs (141). She developed confidence in her own observations: apropos of the swallows' habit of mobbing hawks she noted, "I should have been somewhat skeptical on the subject, had I not myself been an eye-witness to the fact" (187). Considering that during most of the nineteenth century naturalists had no binoculars or field guides but studied animal skins in museums or private collections, and that Catharine Parr Traill had no access to such collections for comparative study, the accuracy of most of her observations is remarkable.

Twentieth-century naturalists may be unlikely to agree with her that the red-winged blackbird's song is "full as fine as our English blackbird," or that this species has "a glossy, changeable, greenish back" (184), a description that fits the much larger common grackle. Yet, she carefully and accurately observed the flocking behavior of the red-winged blackbirds in midsummer and described the warning call that gave them the folk name "harper" (McAtee 59). Catharine Parr Traill was more interested in explaining bird behavior, an area that became popular in the late nineteenth century, than in cataloguing minute plumage variations, or analyzing bird songs.

Birds and animals became peripheral to the theme of *The Female Emigrant's Guide,* which is a pragmatic, "how-to" guide for prospective emigrant families. Nevertheless, in this book Catharine Parr Traill included information about animal behavior and ecological relationships in a month-by-month description of natural events. For January, she wrote, the "bear, the raccoon, the porcupine, the groundhog, the flying squirrel and little striped chitmunk [chipmunk] . . . lie soundly sleeping in their nests or burrows. The woods are deserted by most of the feathered tribe" (212). In May "crowds of birds . . . visit us. The green frogs are piping in the creeks and marshes" (212). In July "the mosquitoes are the most troublesome, especially in the close pine woods, and near lakes and streams. . . . Many splendid coloured butterflies are seen during the hot months . . . " (217). Although the *Guide* was not intended as a popular science book, there are numerous references to practical or applied science, such as mycology (yeast), chemistry (potash and soap making), and nutrition (making jam, selecting apples for drying, preparing apple butter). In fact, Catharine Parr Traill's detailed advice for people interested in agriculture precedes publications by late-nineteenth-century Canadian women applied science writers, such as Eliza M. Jones and Annie L. Jack, as well as Mrs. Beeton's famous English *Book of Household Management.*[4]

Catharine Parr Traill's scientific curiosity led her to enjoy and investigate many aspects of the natural world, but she preferred studying flowers: their medicinal properties, geographic distribution, and life cycle. These interests reflected traditional European studies of plants as *materia medica* and food, late-eighteenth- and early-nineteenth-century concerns about geographic distribution, as well as a new, late-nineteenth-century North American trend: that of long-term life history studies. Three of her major works, *Backwoods, Canadian Wild Flowers,* and *Studies of Plant Life,* written over a fifty-year period, demonstrate her love of science, her willingness to learn from both her own experiences and those of others, her growing knowledge of plant taxonomy and ecology, and her increased concern about habitat destruction.

In *The Backwoods of Canada,* Catharine Parr Traill noted that when she arrived in Upper Canada, in 1832, she had found that "this country opens a wide and fruitful field to the inquiries of the botanist" (80–81). She already had a "passion for flowers" (101), and as she did not know their English, Latin, or Indian names, she did not hesitate to name them. Though aware that "our scientific botanists in Britain would consider me very impertinent in bestowing names on the flowers and plants I met with in these wild woods," in a new country and in the absence of available documentation she felt free, she noted, to "become their floral godmother, and give them names of my own choosing" (120). Thus, quite consciously, she both distanced herself from the European centers of botany and defined herself as a *woman* botanist.

Within a few months, Catharine Parr Traill made dried flower collections, sent plant specimens to Scotland, and studied the size, shape, life cycle, and succession of plants. In less than two years, she learned, partly from firsthand observation and partly from other settlers, that fireweed, sumach, raspberry, wild strawberry, and a variety of grasses and shrubs follow forest fires in a certain sequence. From the First Nations women in the district, she also learned about the medicinal properties of North American plants and about the practical uses of natural products. From *The Backwoods of Canada*, it is also evident that she developed self-confidence from her expanding knowledge. For instance, when she found that Frederick Pursh's *Flora Americae Septentrionalis* (1814) provided no description for a plant, she identified it "to be a species of honey-suckle, from the class and order, the shape and colour of the leaves, the stalks, the trumpet-shaped blossom, and the fruit; all bearing a resemblance to our honeysuckle in some degree" (194).

In the Rice Lake region, where the Traills lived from the late 1840s to the late 1850s, Catharine Parr Traill continued her natural history studies, taking careful notes of plant succession, habitat, and animal behavior. She built up a sizable herbarium, sold more than a dozen natural history articles to the *Anglo-American Magazine* (Toronto) and the *Horticulturalist* (Albany), and wrote other, shorter pieces on plants and animals that were to form integral parts of her later books. In the published articles, her writing was easy and informative: she used familiar and maternal metaphors and scientific terminology interchangeably. In the *Horticulturalist* she wrote about the maidenhair fern that "no trace of its infant dress remain[s] visible on the whole plant. The stem becomes smooth, and black, and elastic, like fine whalebone, supporting its exquisite foliage on foot-stalks of hairlike lightness" (332–33), while the leaves of *Podophyllum peltatum* (mandrake or mayapple)" are folded round the stem like a closed parasol," and its roots are "used by the Indians as a cathartic" (333). She was obviously comfortable with the scientific terms used by taxonomists. As a pioneer settler and mother, living far from doctors and pharmacists, she was interested in the medicinal properties of plants and acknowledged the traditional medicinal knowledge of First Nations women.

Catharine Parr Traill's life was temporarily disrupted in August 1857 when her home, including her dried flower collection and many notes and letters, was destroyed by fire. Resilient and determined, she overcame this loss, but Thomas Traill never recovered. After his death in 1859, the fifty-seven-year-old widow and her younger children returned to the area of their first settlement. With the help of a hundred-pound-sterling literary grant, her widow's pension from England, and her own earnings, she was able to spend more time on her writing. She also developed better scientific, literary, and publishing connections in an increasingly populated Canada. With some difficulty she obtained books of North American as well as British scientists, including *Wildflowers of Nova Scotia,* illustrated by Maria Morris (1840), with very brief explanatory texts by Titus Smith. After 1860, she read more up-to-date science books and journals, but like most other Canadian naturalists she was not influenced by Charles Darwin. John Macoun had "no time" for Darwin's theories, and the geologist and institution builder J. W. Dawson was a vocal opponent of Darwinian evolution. Both Catharine Parr Traill and John Macoun knew, however, the works of American botanist Asa Gray,[5] who was an outspoken Darwinian! That she was not unaware of the hotly debated issues of heredity and evolution can be seen from a letter she wrote to the novelist William Kirby in 1895: "I do believe something in this now fashionable doctrine of heredity."[6]

New or improved roads and better railway connections enabled her to travel periodically to the growing cities of Kingston, Toronto, and Ottawa, where she met with and obtained the support of scientists. This group included the few Canadian professors who by that time were teaching natural history and botany, such as the Reverend William Hincks in Toronto, and Professor George Lawson at Kingston, who recognized her serious interest in natural history. Botany, never a mere hobby for her, by 1860 became her major scientific interest.

Catharine Parr Traill's long-term studies resulted in *Canadian Wild Flowers* (1868) and *Studies of Plant Life in Canada* (1885). The impetus to write a plant book had occurred soon after she moved to Canada in 1832; there were simply no usable botany manuals for her new environment. At the time, she would recall in *Canadian Wild Flowers* thirty-six years later, "there was no one to give written description, or to compile a native Flora, or even a domestic Herbal of the Wild Plants of Canada" (8). She meant to remedy the situation, but three decades passed before she could devote herself to the writing of such a book. In the early 1860s, her manuscript "received the sanction and approval of several scientific and literary gentleman in Canada, among whom were Dr. Hincks and Prof. George Lawson" (8), but Canadian publishers were unwilling to invest in a manuscript, no matter how scientific, that had no illustrations and was therefore unlikely to appeal to the general public. Catharine Parr Traill had never learned flower painting in England, but her Canadian-born niece, the enterprising Agnes Moodie FitzGibbon (later Mrs. Chamberlin) (1833–1913), was willing to illustrate the work to make it an "entirely Canadian production" (Chamberlin). *Canadian Wild Flowers* became the first Canadian botany book with easily accessible scientific text. Pursh's volume (1814), in Latin, was far too technical to reach a broad-based readership. Maria Morris's *Nova Scotia Wildflowers* contained minimal text, and W. J. Hooker's *Flora Boreali Americana* was too expensive and technical to appeal to the general reader. In addition, *Canadian Wild Flowers* was the *first* natural history volume in North America both illustrated and written by women.

The volume contained ten plates of thirty-one species of Canadian wildflowers. Plant groups were arranged more or less in the seasonal order the botanist might encounter them, rather than alphabetically, as in the old herbals, or according to taxonomic sequence, as in botany books written by museum scientists. In her lively, easily readable descriptions, Catharine Parr Traill used technical terms, imparted scientific information, and evaluated the statements of American botanists. Her text for the painted cup is a good example:

> This splendidly-coloured plant is the glory and ornament of the plain-lands of Canada. The whole plant is a glow of scarlet, varying from pale flame-colour to the most vivid vermillion. . . . The flower is a flattened tube, bordered with bright red, and edged with golden yellow. Stamens, four; pistil, one; projecting beyond the tube of the calix; the capsule is many seeded. . . . The American botanists speak of *Castelleia coccinea,* as being addicted to a low, wettish soil, but it is not so with our Canadian plant; . . . it is neither to be found in swamps nor in the shade of the uncleared forest. (15–16)

Because she wrote about plant succession as well as the interrelation of soil, light, climate, plants, and animals, Catharine Parr Traill may be considered as an early ecologist. For instance, she wrote that the "Spring Beauty . . . comes with the Robin, and the Song Sparrow, the hepatica, and the first white violet" (85). And while as a good evangelical Anglican she praised the creator for his "wise and merciful" arrangements, she made sound scientific observations on the "connection between BEES and FLOWERS. In cold climates the former lie torpid, or nearly so during the long months of Winter, until the genial rays of the sun and light have quickened vegetation into activity, and buds and blossoms open, containing nutriment necessary for this busy insect tribe. The BEES seem made for the Blossoms; the BLOSSOMS for the BEES" (85).

As in her earlier articles, in *Canadian Wild Flowers* Catharine Parr Traill wrote about the medicinal properties of plants, the "healing qualities of the large white Lily roots and leaves . . . [to be used as] a poultice to sores and boils" (54). She noted that "The Indian herbalists use the Indian Turnip in medicine as a remedy in violent colic" (10), and that many "of these compound flowers possess medicinal qualities. Some as the thistle, dandelion, wild lettuce . . . are narcotic" (41). As before, she employed maternal metaphors: "on the approach of night our lovely water-nymph gradually closes her petals and slowly retires to rest within her watery bed, to rise on the following day, to court the warmth and light so necessary for the perfection of the embryo seed" (69).

From *Canadian Wild Flowers* it is clear that although Catharine Parr Traill lived far from major centers with their institutes, colleges, and museums, she was familiar with several systems of classification, had studied plant structure through a "powerful microscope" (25), and had acquired considerable field experience. Like many of her contemporaries, she saw nothing contradictory in writing about her own observations of God's arrangements, while classifying plants according to Antoine Laurent de Jussieu's "natural order." Throughout the text she made numerous references to male scientists, such as Hincks, Lawson, Linnaeus, and Jussieu, but was not afraid to offer dissenting views. She wrote, for instance, that "Gray and other botanical writers call this striking flower (T. erèctum) the 'Purple Trillium;' it should rather be called Red, its hue being decidedly more *red* than purple . . . " (39). She also noted that in "the Linnaean classification they were included in common, with all the Orchis tribe, in the class Gynaudria, but in the Natural Order of Jussieu, which we have followed, the [yellow] 'Lady's Slippers' (*Cypripèdium*) form one of the sub-orders in the General Order Orchidaceae" (45). She knew about variation due to geographic distribution and the possibility for making new botanical discoveries "native to Canadian soil" (65). When it suited her, she relied on the scientific authority of men to legitimize her claims. Thus, her description of eelgrass "behaviour" was "vouched for by Dr. Gray and many other creditable botanists" (69). Catharine Parr Traill was writing for a mixed audience: other naturalists as well as the general public. She referred to Asa Gray and "other creditable botanists" because, in spite of her own private training in England and field experience on both sides of the Atlantic, she remained conscious about her lack of museum training in systematics/taxonomy.

In 1868 *Canadian Wild Flowers* was the only widely accessible scientific botany book on Canadian plants. Catharine Parr Traill's lively and informative descriptions, combined with Agnes FitzGibbon's illustrations, ensured that the book was well received by reviewers and the public. The 23 June 1875 issue of the *New Century* referred to it as "One of the most remarkable works ever attempted by a woman." The *Montreal Daily News* wrote, "Mrs. Traill, in simple and beautiful language well suited to the subject . . . describes these wildings of the woods with the love and enthusiasm of a student of nature." A "Boston paper" referred to it as "a volume that will be as memorable as that of Madame [Maria Sybilla] Merian" (Briggs).

After *Canadian Wild Flowers* was reprinted several times, Catharine Parr Traill was encouraged to write a more complete volume. But the work progressed slowly, and by the time *Studies of Plant Life in Canada* was published (1885) the status of botany had changed in Canada. During the 1870s and early 1880s, botany became institutionalized. Professor John Macoun (1831–1920), another backwoods settler and self-trained botanist, worked on government surveys of western Canada, made large plant collections, and published a list of plants in 1878. After he was hired in 1882 as Dominion Botanist by the Geological and Natural History Survey of Canada, he published *Catalogue of Canadian Plants* (1883–86). According to W. A. Waiser, the three-part, 608-page work provided "the synonymy, habitats, and collections for every known species of . . . plant in Canada" (80). It was useful but appealed to a very limited number of naturalists. This reinforced Catharine Parr Traill's wish to publish a readable volume that could appeal to both the general public and the naturalist.

Studies of Plant Life was edited, as a favor, by James Fletcher, Parliamentary Librarian and a keen naturalist, who, after the establishment of the Central Experimental Farm in Ottawa in 1886, was to become Dominion Entomologist of Canada. The extensive correspondence between Catharine Parr Traill and James Fletcher indicates his willingness to help her and demonstrates the active exchange of ideas and specimens between the two. In fact, well into her nineties, Catharine Parr Traill continued to send mosses, ferns, lichens, and fungi for the government's botanical collection and Fletcher's private one.

Studies of Plant Life in Canada is no dry, scientific catalogue but a book about the flowers, shrubs, trees, and ferns Catharine Parr Traill had observed and studied during a fifty-year period. In the preface the author stated that this was not a book for the

> learned. . . . The writer has adopted a familiar style in her description of the plants, thinking it might prove more useful and interesting to the general reader, especially to the young, and thus find a place on the book-shelves of

many who would only regard it for the sake of its being a pretty, attractive volume on account of the illustrations [by Agnes Chamberlin].(i)

She urged the "Mothers of Canada [to] teach your children to know and love the wild flowers springing in their path, to love the soil in which God's hand has planted them"(ii). At a time when "civilization" was well on its way to destroy the "forest land and prairie, . . . the native trees and the plants that are sheltered by them"(ii), Catharine Parr Traill planned to preserve "their beauties" on the written page (ii).

Studies is divided into four parts: native or wild flowers, flowering shrubs, forest trees, and ferns. As before, she arranged the plants "in the order in which they appear in the woods," though in some cases, especially with the short species descriptions, she also grouped them "somewhat in families" (iii). *Studies* is three times longer than *Canadian Wild Flowers:* there is more information on geographic variation, ecological relationships, medicinal and other uses of plants, and practical uses of plant dye. From this book it is clear not only that Catharine Parr Traill expanded her scientific knowledge but that she had become an active participant in the informal exchange of North American plants. There are references to specimens received from other parts of the continent, some sent by relatives, others by unnamed naturalists in the United States, the Northwest Territories, and Lower Canada.

Because the author planned this book for the general public, but especially for "Canadian mothers," she used less scientific terminology than either in her previous botanical articles or in *Canadian Wild Flowers.* She provided, however, English, scientific, and native names for the plants she described, and again included useful information gathered from First Nations women and old settlers. Unlike most other science writers in the colonies, Catharine Parr Traill integrated rather than marginalized nonwestern scientific information and practices. She wrote, for instance, that the "orange fibrous roots and rootlets" of the gold thread are "intensely bitter, and are much used by the old settlers as tonic remedies. . . . The Indian women use it for their little ones in case of a sore mouth and sore gums in teething" (37–38). The "leaves of this beautiful Wintergreen are held in high estimation by the Indian herbalists who call it Rheumatism weed (*Pipsissewa*)"(48). About the Canadian balsam she wrote: "The Indian women use the juice in dyeing, and also apply it" for skin problems, such as those "caused by Poison Ivy"(59). As before, the author used maternal metaphors; she referred to the "life-containing germ" of a simple fern as "a sort of maternal instinct . . . imparted to . . . shield the embryo from every possible injury and to insure its safety through all the mysteries of its infant state" (254).

Catharine Parr Traill's concern for future generations of young Canadians prompted her to express her views on disappearing plants and animals and on the need for their preservation (151–53). She explored issues that concerned few Canadians in the late nineteenth century, such as the need to preserve fragile habitats and create national parks (212–13). In this, as in many of her studies, she was in the forefront of natural history and conservation in Canada (Foster).

Throughout her long life, Catharine Parr Traill was well-known in England as a writer and a botanist. In North America too, her botanical and natural history studies were read by men, women, and children of all classes. She remained ambivalent, however, about being called a botanist. She referred to museum botanists and/or taxonomists, as "botanists" and criticized them while demonstrating her own knowledge of botanical terms. Her article "Lilies," published in the *Horticulturalist* in 1853, is a good example:

> Botanists seem to me fond of separating the members of this fair family, and putting asunder those who nature has joined together. All bulbous-rooted, hexandrous, hexapetalous flowers, are naturally allied and should be, I think, classed in one order. I would arrange all the families of plants in grades, or steps, linking them together in a great natural chain. Botanists would doubtless think me presumptuous in proposing any classification opposed to the popular ones in vogue, and possibly great objections might exist which I have not considered. I therefore offer the suggestion with all humility and deference to the learned in the science [i.e., botany]. (520)

Paying lip service to how a mid-nineteenth-century Englishwoman should think and act, Catharine Parr Traill both challenged the dominant western scientific terminology and took a sly dig at taxonomists, who, before the trinomial nomenclature was accepted in the late nineteenth century, were known to turn each new variety into a full species. Interestingly, while she referred to early-nineteenth-century scientific notions about the great chain of being, her view also presaged the late-twentieth-century concept of clines.[7] Her familiarity with new works on botany is evident, as when she quotes not only "Sir James Smith (a name celebrated among English botanists)" (520) but also Dr. John Richardson (who provided many of the specimens for *Flora Boreali Americana* by W. J. Hooker).

Although she remained an active botanist and science writer to the end of her long life, Catharine Parr Traill was aware of the changing practices of science, the increased emphasis on laboratory work, and the growth of professionalization. In 1894, at age ninety-two, she still resisted being called a "botanist," to which James Fletcher responded: "With regard to your disclaiming the title of botanist . . . I wish that a fraction of one per cent of the students of plants who call themselves botanist, could use their eyes half as well as you have done. I think indeed your work of describing all the wild plants in your book, so accurately that each could have the name applied to it without doubt, is one of the greatest botanical triumphs which anyone could achieve. . . ."[8]

Catharine Parr Traill most certainly was not a "splendid anachronism," nor was her work that of a struggling "amateur." Her botanical work was similar to that of American women science writers, such as Almira H. L. Phelps (1793–1884), Graceanna Lewis (1821–1912), and Florence Merriam Bailey (1863–1948), although there is no evidence that she ever read their works.[9] Unlike most of her British counterparts, but like many American women naturalists, she had the opportunity to explore new areas, observe geographical differences in plant and animal distribution, and "discover" new plants.[10] More important, she challenged the paternalistic system of western science.

Although her initial training was in the British natural history tradition of Gilbert White, Catharine Parr Traill's scientific explorations of the Canadian flora and fauna fit well the emerging North American trend of long-term life-history studies which, by the last quarter of the nineteenth century, were to complement and to some extent supersede the collection-based study of taxonomy. Her botanical investigations, though restricted to a limited area, grew out of her own field experience. In popularity and accessibility *Canadian Wild Flowers* and *Studies of Plant Life* can be considered the forerunners of the vastly popular twentieth-century phenomenon, the *field guide*. In *Canadian Crusoes*, she wrote about survival techniques in the Canadian "wilderness" fifty years before Ernest Thompson Seton wrote about wood lore, though in her book the environmental knowledge is communicated by a young native woman rather than by boys and men. According to Caitling, in the 1990s her botanical writing provides an outstanding historical record of the "extent and floristic composition" of the Rice Lake Plains. Through her, we can still learn about habitat destruction, changes in plant and animal life, ecological succession, and native environmental knowledge and beliefs. Her writing provides a postcolonial alternative to earlier western scientific texts.

Catharine Parr Traill (1802–1899): Major Works

The Backwoods of Canada: Being Letters from the Wife of an Emigrant Officer, Illustrative of the Domestic Economy of British North America (1836)

Canadian Crusoes: Tales of the Rice Lake Plains (1852)
The Female Emigrant's Guide, and Hints on Canadian Housekeeping (1854)
Canadian Wild Flowers (1868)
Studies of Plant Life in Canada; or, Gleaning from Forest, Lake, and Plain (1885)
Cot and Cradle Stories (1895)

Notes

I would like to thank Ann B. Shteir and Barbara Gates for encouraging me to undertake this project, Ariel Fielding for her careful and insightful research, Sally Cole and Elizabeth Parnis for providing valuable information, and David Ainley, Karin Beeler, Tina Crossfield, Susan Hoecker-Drysdale, and Barbara Meadowcroft for their helpful comments and discussions.

1. Many twentieth-century authors use the term "amateur" in a pejorative sense meaning dilettante, rather than in the original sense of "amator" meaning "lover of." Catharine Parr Traill was most certainly a lover of nature. She was also a very competent naturalist at a time when science in Canada was far from professionalized. On the amateur-professional continuum in science, see for instance Stebbins; Ainley, "Contribution."

2. Suzanne Zeller, pers. com.

3. Similar colonial situations existed in other British and European colonies, e.g., Australia. Though earlier works by Lewis Pyenson and Roy MacLeod on colonial science excluded gender, it is puzzling to see such gender blindness in a more recent collection, Macleod and Jarrell. Fortunately, there is a rapidly growing postcolonial literature on women in Australia, South Africa, and the Americas. See for instance Maroske.

4. On Jack and Jones see Ainley, "Women and the Popularization of Science." In her introduction to the New Canadian Library edition of the *Canadian Settler's Guide* (1969), Clara Thomas argues that Catharine Traill "was certainly the Mrs. Beeton of nineteenth century Canada." But, "in comparison to Mrs. Beeton, Mrs. Traill offers woman, and demands of her, a great range of adaptability, competence, endurance and rewards" (x).

5. See Susanna Moodie to C. P. Traill, 30 July 1856, Traill Family Papers, National Archives of Canada (hereafter TFP).

6. C. P. Traill to Wm. Kirby, 20 February 1895, Lorne Pierce Papers, Queen's University Archives.

7. A "cline is a continuous gradation of form differences in a population of a species, correlated with its geographical or ecological distribution" (*Dictionary of Biology* 66).

8. James Fletcher to C. P. Traill, 28 July 1894, TFP.

9. Almira Hart Lincoln (later Phelps), *Familiar Lectures on Botany* (1829), was the first book on North American botany that John Macoun read (Macoun). Graceanna Lewis was a Philadelphia Quaker, a well-known naturalist and ornithologist; her book *The Natural History of Birds* was published in 1868. Her articles on natural history appeared in *The American Naturalist* and

The New Century. Florence Merriam Bailey was a conservationist and ornithologist who wrote popular books as well as a scientific monograph and numerous scientific articles.

10. A local variant of the marginal shield fern was named after her by Professor George Lawson of Queen's University. Elizabeth Parnis, pers. com.

Works Cited

Ainley, M. G. "The Contribution of the Amateur to North American Ornithology: A Historical Perspective." *Living Bird* 18 (1979–80): 161–77.

Ainley, M. G. "From Natural History to Avian Biology: Canadian Ornithology, 1860–1950." Diss., McGill U, 1985.

Ainley, M. G. "Last in the Field?: Canadian Women Natural Scientists, 1815–1965." In *Despite the Odds: Essays on Canadian Women and Science*, ed. M. G. Ainley. Montreal: Véhicule P, 1990. 25–62.

Ainley, M. G. "Women and the Popularization of Science: Nineteenth Century Women Science Writers in Canada." Paper presented at the 8th Kingston Conference of the Canadian Science and Technology Historical Association, Kingston, Ontario, October 1993.

Ballstadt, Carl A. "Catharine Parr Traill (1802–1899)." In *Canadian Writers and Their Works*, ed. Robert Lecker et al. Downsview, Canada: ECW, 1983. 149–93.

Basalla, G. "The Spread of Western Science." *Science* 156 (1967): 611–21.

Berger, Carl. *Science, God, and Nature in Victorian Canada*. Toronto: U of Toronto P, 1983.

Briggs, William. Press Reviews, Traill Family Papers, # 12843–46.

Caitling, P. M, V. R. Caitling, and S. M. McKay-Kuja. "The Extent, Floristic Composition, and Maintenance of the Rice Lake Plains, Ontario, Based on Historical Records." *Canadian Field-Naturalist* 106 (1992): 73–86.

Chamberlin, Agnes. "Introductory Notes to Catharine Parr Traill." *Canadian Wild Flowers*. 4th ed. Toronto: William Briggs, 1895.

Chartrand, Luc, Raymond Duchesne, Yves Gingras. *Histoire des sciences au Québec*. Montreal: Boreal, 1987.

Cole, Jean M. "Catharine Parr Traill—Botanist." *Portraits: Peterborough Area Women Past and Present*. Peterborough, Ont.: Portrait Group, 1975. 73–79.

Croft, L. R. "P. H. Gosse in Newfoundland and Lower Canada, 1827–1838." *Archives of Natural History* 20 (1993): 333–47.

A Dictionary of Biology (Harmondsworth, Middlesex: Penguin Books, 1973).

Dictionary of Canadian Biography. 1990. 12: 995–99.

Eaton, Sara. *Lady of the Backwoods: A Biography of Catharine Parr Traill*. Toronto: Hogger and Stoughton, 1969.

Farber, Paul L. *The Emergence of Ornithology as a Scientific Discipline, 1760–1850*. Dordrecht: D. Reidel, 1982.

Fitzgibbon, Mary Agnes. "Biographical Sketch." *Pearls and Pebbles; or, Notes of an Old Naturalist*. Toronto: William Briggs, 1894. i–xxxiv.

Foster, Janet. *Working for Wildlife: The Beginning of Preservation in Canada.* Toronto: U of Toronto P, 1978.

Fowler, Marian. *The Embroidered Tent: Five Gentlewomen in Early Canada.* Toronto: House of Anansi, 1982.

Levere, Trevor H. *Science and the Canadian Arctic: A Century of Exploration, 1818–1918.* Cambridge: Cambridge UP, 1992.

MacCallum, Elizabeth. "Catharine Parr Traill, A Nineteenth-Century Ontario Naturalist." *Beaver* 1975 (Autumn): 39–45.

MacLeod, Roy, and Richard Jarrell, eds. "Dominions Apart: Reflections on the Culture of Science and Technology in Canada and Australia, 1850–1945." *Scientia Canadensis* 17, nos. 1 and 2 (1994).

Macoun, John. *The Autobiography of John Macoun: Canadian Explorer and Naturalist, 1831–1920.* 2d ed. Ottawa: Ottawa Field-Naturalists Club, 1979.

McAtee, W. L. *Folk Names of Canadian Birds.* Bulletin 149. Ottawa: National Museum of Canada, 1957.

Maroske, Sara. "The Whole Great Continent as a Present: Nineteenth-Century Australian Women Workers in Science." In *On the Edge of Discovery: Australian Women in Science,* ed. Farley Kelly. Melbourne: Melborne UP, 1993. 13–34.

Morris, Audrey Y. *Gentle Pioneers: Five Nineteenth-Century Canadians.* Toronto: McClelland and Stewart, 1968.

Needler, G. H. "The Otonabee Trio of Women Naturalists: Mrs. Stewart, Mrs. Traill, Mrs. Moodie." *Canadian Field-Naturalist* 60 (1946): 97–101.

Peterman, Michael A. " 'A Splendid Anachronism': The Record of Catharine Parr Traill's Struggles as an Amateur Botanist in Nineteenth-Century Canada." In *Re(dis)covering Our Foremothers: Nineteenth-Century Canadian Women Writers,* ed. Lorraine McMullen. Ottawa: Carleton UP, 1990. 173–85.

Pringle, James S. "Anne Mary Perceval (1790–1876), An Early Botanical Collector in Lower Canada." *Canadian Horticultural History* 1 (1985): 7–13.

Stebbins, R. A. "The Amateur: Two Sociological Definitions." *Pacific Sociological Review* 20 (1977): 583–605.

Traill, Catharine Parr. *The Backwoods of Canada: Being Letters from the Wife of an Emigrant Officer, Illustrative of the Domestic Economy of British North America.* London: Charles Knight, 1836.

Traill, Catharine Parr. *Canadian Crusoes: Tales of the Rice Lake Plains.* Ed. Agnes Strickland. London: Arthur, Hall, Virtue, 1852.

Traill, Catharine Parr. *Canadian Wild Flowers.* Montreal: John Lovell, 1868.

Traill, Catharine Parr. *Cot and Cradle Stories.* Ed. Mary Agnes Fitzgibbon. Toronto: Briggs, 1895.

Traill, Catharine Parr. *The Female Emigrant's Guide, and Hints on Canadian Housekeeping.* Toronto: Maclear, 1854. Reprinted as *The Canadian Settler's Guide,* Toronto: Old Countryman's Office, 1855.

Traill, Catharine Parr. *Lady Mary and Her Nurse; or, A Peep into the Canadian Forest.* London: Arthur, Hall, Virtue, 1856.

Traill, Catharine Parr. *Pearls and Pebbles; or, Notes of an Old Naturalist.* Toronto: Briggs, 1894.

Traill, Catharine Parr. *Studies of Plant Life in Canada; or, Gleaning from Forest, Lake, and Plain*. Ottawa: A. S. Woodburn, 1885.
Traill, Catharine Parr. *The Young Emigrants; or, Pictures of Life in Canada, Calculated to Amuse and Instruct the Minds of Youths*. London: Harvey and Darton, 1826.
Waiser, W. A. *The Field Naturalist: John Macoun, the Geological Survey, and Natural Science*. Toronto: U of Toronto P, 1989.
Zeller, Suzanne. *Inventing Canada: Early Victorian Science and the Idea of a Transcontinental Nation*. Toronto: U of Toronto P, 1987.

6

The "Very Poetry of Frogs"
Louisa Anne Meredith in Australia

Judith Johnston

In the early years of the nineteenth century, a young woman called Louisa Anne Twamley received an ordinary English education. She had a daily governess who taught her English, French, and music, and later she was instructed in drawing and sketching, and showed some considerable skill. In her early teens she had the good fortune to receive tuition on the painting of miniatures on ivory from the distinguished artist Thomas Lawrence, whose sister was married to Louisa's uncle (Rae-Ellis 29–33). While botany did not figure in her formal education, nevertheless, in the opening years of the Victorian era, Twamley produced a series of studies on English flowers, poetical in appearance, but often botanical in intent. These books were copiously illustrated by reference mostly to traditional, but also to some modern, English poetry, and with colored plates taken from her own sketches.

The third edition of *The Romance of Nature or, the Flower Seasons Illustrated*, under her newly married name, Mrs. Louisa Anne Meredith, appeared just before she left for Australia in 1839, and although she claims in the preface not to be conveying scientific information because the work is "purely poetical" and "because my own knowledge of botany is too limited to allow of my offering any instruction to others" (viii), nevertheless in a discussion of the forget-me-not she claims to describe the "real" flower, "because other species are continually being

Figure 6.1. Louisa Anne Meredith, née Twamley (1812–1895). Photograph probably taken in the 1860s and reproduced from a *carte de visite.* Allport Library and Museum of Fine Arts, State Library of Tasmania, Tasmania, Australia.

mistaken for the true one." She adds: "Among other instances of this, the illustrator of a recent serious work on Flowers, although professedly a botanical draughtsman, gives the *Myosotis Alpestris* instead of the *Myosotis Palustris,* and so exaggerates the hairy surface of the leaves that they seem equipped in winter clothing" (157).

Thus, despite Meredith's prefatory disclaimer, she is obviously prepared to offer "scientific" instruction after all. She also provides an interesting example of the kind of physical and social constraint that women faced in even the most ordinary kind of botanizing in describing with undisguised triumph her success in obtaining a specimen of gorse to paint:

> a party of most correct looking promenaders passed along the road below me, and hearing a rustling in the leaves, looked up with no small astonishment on beholding a figure, they were accustomed to see walking in the town with infinite staidness and propriety, perched up at a height that implied a necessity for most resolute scrambling. My amusement far exceeded their surprise; but I have no doubt my flower-love in this instance gained me the character of a most uncouth young person. (248)

In *The Romance of Nature,* Meredith notes that her enthusiasm is "little aided by scientific knowledge as yet" and anticipates the time when she will be more knowledgeable about "the fascinating study of Botany" (235). By the time Meredith publishes *Notes and Sketches of New South Wales* in 1844, while she is once again disclaiming any ability to convey information "to those skilled in scientific detail" and asserting that her work is "devoid of scientific lore" (vii), the very use and repetition of the term "scientific" suggest that she writes with greater consciousness of a readership seeking more precise, detailed descriptions than her earlier almost naive effusions contained, descriptions that will satisfy both literary and scientific demands.

Simon Gikandi in his article "Englishness, Travel and Theory: Writing the West Indies in the Nineteenth Century" suggests that because the traveler can claim difference through "facticity and the authority of the eyewitness," travel is "more than a sentimental journey to the reaches of empire; embedded within the new science of natural history, it is driven by an ethnographic mission, what Fabian has called a project of 'observation, collection and classification, and description' " (52). The "new science of natural history" became available both to women and to men, although dominated professionally by men. While Lucile Brockway in *Science and Colonial Expansion* writes that the "voluntary associations of British professionals and amateurs devoted to science were a significant force in the generation and spread of new knowledge" (72), Ann Shteir in her article "Linnaeus's Daughters: Women and British Botany" points out that women were cultivators of science, because they "stimulated an interest in exploring nature, enlarged the data base, and spread botanical knowledge"; they also encouraged book publication (69). At the same time, however, botanical study for women also served as "a form of social control" (73) because it was viewed as a harmless activity.[1]

From Australia the *Tasmanian Journal of Natural Science* first appeared in 1842 and was reviewed in the London *Monthly Review* the same year. The reviewer notes particularly that "as colonization enlarges the territory of any civilized people, so also, as a necessary consequence, will the empire of science be widened and cultivated" ("Tasmanian Journal" 520). Surprisingly, the reviewer recommends the study of nature to men in terms similar to those used when nature study is recommended to women, as "peculiarly desirable and necessary when far removed from the stir of Europe, and cast upon a shore where monotony and seclusion predominate." Nature study will preserve the mental faculties from being eroded "through inactivity, or, what is worse, not only narrowed by exclusive converse with petty details, but debased by sordid, perhaps gross, passions" ("Tasmanian Journal" 523).

Certainly English publications on natural history and botany abounded in the 1830s and 1840s. A typical scientific publication of the period was the *Penny Cyclopaedia of the Society for the Diffusion of Useful Knowledge,* published in 1838, which has an entry on frogs alone containing almost ten pages of detailed physical and biological description, including woodcuts of frog skeletons and the recounting of horrific experiments. The prose style is straightforward, lucid, and unadorned: "They have no ribs, or rudiments of ribs only. Their skin is naked, being without scales; they have feet. The male has no external organ of generation, and there is consequently no intromissive coïtus" ("Frogs, Frog-tribe" 486).

The Zoology of the Voyage of the Beagle, "Part V, Reptiles," was written by Thomas Bell and published in 1843. This work appears to be aimed at the more educated naturalist. The only concession to a literary style is exampled in Bell's physical description of a frog "first discovered by Messrs. Quoy and Gaimard at King George's Sound, in Australia, where it was also obtained by Mr. Darwin" (34). Bell continues: "It is a beautiful species; the back being of a rich brown colour, with a pale orange fascia extending along the sides from the eye to the thigh, becoming bright orange on the flanks. Thighs and legs banded with rich deep brown and bright orange" (34). Only the modifiers "beautiful," "rich," "pale," and "bright" are literary.

Robert Chambers published *Vestiges of the Natural History of Creation* in 1844, a scientific work that captured not only the imagination of the naturalists but the poetic imagination too. Tennyson's *In Memoriam* (1850) reveals the poet incidentally reflecting the scientific tone of the age in showing the influence of both Chambers's *Vestiges* and Lyall's *Principles of Geology.* Chambers's literary style is quite pronouncedly different both from that of Bell just cited, and from that of the quotation from the *Penny Cyclopaedia:*

> It must be borne in mind that the gestation of a single organism is the work of but a few days, weeks, or months; but the gestation (so to speak) of a whole creation is a matter probably involving enormous spaces of time. Suppose that an ephemeron, hovering over a pool for its one April day of life, were capable of observing the fry of the frog in the water below. In its aged afternoon, having seen no change upon them for such a long time, it would be little qualified to conceive that the external branchiae of these creatures were to decay, and be replaced by internal lungs, that feet were to be developed, the tail erased, and the animal then to become a denizen of the land. (Chambers 210–11)

Ann Shteir has written that women were constrained by both geographical and physical barriers in the eighteenth and nineteenth centuries, as Louisa Meredith's anecdote about scrambling for a specimen of gorse indicates. Only colonizing women had the opportunity to botanize in a more exotic terrain than the British woods or seashore could offer (72). Those who would have been serious professional botanists were also constrained by insurmountable social barriers. In Britain women were steadfastly refused entry to the leading scientific societies, the Linnaean Society admitting women in 1919, the Royal Society not until 1946 (Shteir 68).

Louisa Anne Meredith became one such colonizing woman. In 1845 she published *Notes and Sketches of New South Wales.* She forwarded the completed manuscript from Tasmania in December 1843 (Rae-Ellis 122) to John Murray in London, who published the work in their "Colonial and Home Library" series. Her book was popular and ran to four editions in the next seven years (Rae-Ellis 150). That popularity was no doubt in part due to the fact that Meredith had already published several well-reviewed volumes of botanical studies before her marriage and subsequent removal to Australia, and in part a result of excellent reviews of the work on New South Wales in some of the most influential journals of the day, among them the *Quarterly Review,* the *Athenaeum* and *Tait's Edinburgh Magazine.*

Elizabeth Rigby reviewed Meredith's *Notes and Sketches,* with twelve other books, in her lengthy article titled "Lady Travellers," published in the *Quarterly Review* in 1845.[2] Rigby, herself a highly successful writer and journalist, mediates Meredith's text to a large potential readership, claiming in her opening remarks that a woman's travel writing is "all ease, animation, vivacity, with the tact to dwell upon what you most want to know, and the sense to pass over what she does not know herself" (99). Women's travel writing, Rigby continues, always unconsciously reveals the writer, and the Englishwoman in particular:

> with her national courage and her national reserve, with her sound head and her tender heart, with the independent freedom of her actions and the decorous restraint of her manners, with her high intellectual acquirements and her simplicity of tastes, . . . versed in the humblest indoor duty, excelling in the hardiest out-door exercise; . . . enthusiastic for nature; keen for adventure. (100)

According to Rigby, the Englishwoman "excels all others in the art of travelling"; no "foreign" woman (she cites particularly women from Germany and France) ever combines the "four cardinal virtues of travelling—activity, punctuality, courage, and independence—like the Englishwoman" (102). Rigby disseminates her own brand of individual prejudice, asking "where is she whose comforts are nine-tenths of them comprised under the head of fresh air and plenty of water, like the Englishwoman's?" (102). Rigby effectively produces a "privileged English identity," to use Simon Gikandi's phrase, in which "Englishness is always the stable point of reference" (50–51). At the same time, this identity in Rigby's text is specifically gendered female. Louisa Anne Meredith is singled out by Rigby as illustrating "the distinctive traits" of her representative paragon, the English woman writer. Meredith has "the easy style—the brilliant thought—the delicate touch—the close detail—the sound sense . . . which gives the true healthy English tone" (105–6).

For Rigby, however, the major attraction that Meredith's book holds is not that of a travel text, but rather the

> valuable store of natural history it communicates. Under a name she has since changed—we think for the better—this lady is well known to the flower-loving world as the most graceful expositor of English botany; and this volume proves that her taste and knowledge extend to many other departments of natural phenomena. Birds and beasts, fishes and insects, and creeping things innumerable equally engage her intelligent attention, and are described with a simplicity and precision which will give much valuable information to the professed naturalist, no additional jargon to the dabbling amateur, and involuntary interest to the most uninitiated. (106)

Rigby's review thus establishes for Meredith a privileged English identity that is also specifically female and literary, and moreover validates her nature study as scientific by proclaiming its value to the amateur and the professional alike.

Louisa Meredith's *Notes and Sketches of New South Wales,* however, is not strictly a book of nature study, but indeed is a travel text, with that shifting from topic to topic that simulates the moving from place to place of the traveler. Chapter 11, for instance, details a sojourn in Bathurst in which we discover studies of local fauna juxtaposed with observations on, and descriptions of, indigenous people.[3] Some reviewers highlighted Meredith's accounts of local people. *Hogg's Weekly Instructor,* for instance, reprinted a section and titled it "Aborigines of New South Wales," and Henry Chorley, reviewing her work for the *Athenaeum,* similarly featured this section.

Elizabeth Rigby, however, makes no mention at all of Meredith's accounts of local people. Instead, her review celebrates and focuses upon Meredith's nature study, to the exclusion of most other topics. One particular passage indeed is introduced as "the very poetry of frogs":

> In the Macquarie, near Bathurst, I first saw the superb green frogs of Australia. The river, at the period of our visit, was for the most part a dry bed, with small pools in the deeper holes, and in these, among the few slimy water-plants and Confervae, dwelt these gorgeous reptiles. In form and size they resemble a very large common English frog; but their colour is more beautiful than words can describe. I never saw plant or gem of so bright tints. A vivid yellow-green seems the groundwork of the creature's array, and this is daintily pencilled over with other shades, emerald, olive, and blue greens, with a few delicate markings of bright yellow, like an embroidery in threads of gold on shaded velvet. And the creatures sit looking at you from their moist, floating bowers, with their large eyes expressing the most perfect enjoyment, which, if you doubt whilst they sit still, you cannot refuse to believe in when you see them flop into the delicious cool water, and go slowly stretching their long green legs, as they pass along the waving grove of sedgy, feathery plants in the river's bed, and you lose them under a dense mass of gently waving leaves; and to see this, whilst a burning, broiling sun is scorching up your very life, and the glare of herbless earth dazzles your agonized eyes into blindness, is almost enough to make one willing to forego all the glories of humanity, and be changed into a frog![4] (*Notes and Sketches* 107)

The poetry of frogs offers no biological detail of the kind found in the *Penny Cyclopaedia*. Meredith's description indeed opposes the *Cyclopaedia* dismissal of the skin of frogs as "naked," by signaling that frogs in Australia are different because costumed in an exotic, colorful garb. Had she been describing people, Meredith's discussion must have proved a classic example of "othering" by her emphasis on a nineteenth-century version of medieval splendor. Sara Mills in *Discourses of Difference* has discussed how feudal or medieval comparisons are a discursive practice meant to relegate those observed back into a more primitive European past (89). Meredith's description of the frogs as clothed is not particularly idiosyncratic either, but appears to be a commonplace of the period. Catharine Parr Traill, in *The Backwoods of Canada,* first published in 1836, writes similarly of frogs in Canada:

> The green frogs are very handsome, being marked over with brown oval shields on the most vivid green coat: they are larger in size than the biggest of our English frogs, and certainly much handsomer in every respect. (141)

The difference in Meredith's writing is that her prose style is both exotic, with such words as "gorgeous," and heroic. Descriptions of the "frog's array" as "daintily pencilled," having "embroidery in threads of gold on shaded velvet," suggest the caparison of medieval knights more particularly than does Traill's reference to "brown oval shields." Meredith's writing takes on a lyric quality that evokes a Victorian medievalism of the kind located in poems like Tennyson's "The Lady of Shalott" (1832). Her "floating bowers" and "waving grove of sedgy, feathery plants" faintly echo the "plumes and lights" of Camelot, and Lancelot riding "between the barley-sheaves," the plumes on his helmet nodding in the sun, dazzling the agonized eyes of the captive Lady of Shalott, just as the Australian sun "dazzles your agonized eyes into blindness."

Meredith paints a word picture that attempts to capture appearance and difference, and if her terms are poetic, even chivalric, nevertheless what she achieves is as exact in its way as that apparently scientific description of Australian frogs by Bell, based on Darwin's notes. Or perhaps the lyricism of Chambers is closer to her style when she expatiates on the "glories of humanity," although in Meredith's text the procedure is reserved, human into frog rather than the other way around. Unlike Bell, Meredith has the added advantage of being an observer on the ground. Her description is very much of creatures *in situ,* and what dominates this passage is a strong sense of place and season. There is no evidence whatsoever of scientific detachment, but a subjective delight in what she records.

Meredith first mentions indigenous people in chapter 10, and dress, or rather undress, is the main focus of her paragraph. She records a scene described to her by her husband—she is therefore no longer an eyewitness but the purveyor merely of another's impressions. She writes: "the sable gentry assembled by degrees as they completed their evening toilettes, *full dress* being painted nudity," and she notes that the "whole 'tableau' is fearfully grand: the dark wild forest scenery around—the bright fire-light gleaming upon the savage and uncouth figures of men" (*Notes and Sketches* 91).

As I have pointed out briefly in a previous article (Johnston 49) Meredith resorts here to a discourse that is reminiscent of the Augustan mode. It is this mode that dominates any description of aboriginal ceremonial in Meredith's first Australian book designed for her Victorian audience "back home"; for instance, the phrase "sable gentry" is part of that false elevation of language, a preposterously inflated lexis, deliberately employed to mock. Subsequently she writes more fully:

> In lieu of "Rowland's inestimable oil Macassar,"[5] their black elvish locks are always plentifully loaded with opossum or snake fat, which unsavoury unguent, as may be imagined, adds its share to the powerful and not too-pleasing odour natural to them.
>
> A sable exquisite preparing for an evening party first undresses, then thrusts a large lump of pipe-clay into his mouth to soften, and when of a proper consistency uses his forefinger as a pencil, dipping it into the composition, and carefully dispensing the cherished ornament over his person. . . . this done, no reigning belle of the season ever entered Almack's with more consciousness of all-powerful beauty than he feels in taking his place among the equally elaborate costumes of his companions. I fear the poor young squaws, or "Gins," have but little to do with their own disposal in marriage; but doubtless many a tender heart must be touched by these D'Orsays of the wilderness. (*Notes and Sketches* 98–99)

At this point in her discourse, and at this early date, the only difference between those she terms the "savage nations" (*Notes and Sketches* 93) and those she terms "civilized" (*Notes and Sketches* 92) is, by George Stocking's definition, simply "the progress of refinement and of civilization itself" (18). Thus Meredith's explicit comparisons with the British upper-class establishment's own elaborate courting rituals from the Regency period, which her references to Byron's *Don Juan*, Almack's, and Count d'Orsay[6] suggest, and her response to personal adornment as feminine rather than masculine, register difference but at the same time mock that difference through a highly stylised and at times antiquated literary discourse. This said, Meredith does qualify her discussion overall by pointing out civilized social practices among the Aboriginal people like reverence for the elderly and the ready acceptance of orphaned children into alternative family groups, as well as noting the variety of languages around the continent.

In *Victorian Anthropology,* George Stocking notes that after the Crystal Palace was moved to Sydenham in 1852, the newly established Natural History Department grouped native people with the "other fauna," eskimos with polar bears and so on (47). The Linnaean system, so popular in botanizing, informed the approach of physical anthropologists who "were interested in classifying the 'types of mankind' rather than in reconstructing their 'physical history' " (Stocking 67), and with the publication of Robert Latham's *Varieties of Man* in 1850, Stocking claims that ethnology "retreated into a purely descriptive 'natural history' of the sort pursued by early eighteenth-century botanists" (103). However, the racialism that was already emerging, and of which there are distinct signs in Meredith's ambivalent prose in her text of 1844, would eventually harden into a determination that non-European races were inferior, impervious to social change, and had no common origin with Europeans. Stocking writes that Darwin's observations of the Fuegians, made in his diary of the voyage of the *Beagle*, "in which paint and grease and body structure blended into a single perception of physical type, perceptually unseparated from what he heard as discordant language and saw as outlandish behaviour," resulted in his subsuming this pattern of experiences under the one term "race" (106). This, Stocking suggests, is quite consistent with the natural historian's treatment of animal species "in which body type, cries or calls, and habitual behaviour were all data to be used in distinguishing a variety or 'race.' " This is manifest in chapter 7 of *The Descent of Man* (1871) where Darwin's imagined or supposed naturalist puts forward an argument "in favour of classing the races of man as distinct species" (Darwin 260). Darwin, it should be noted, puts forward his naturalist's argument in order to oppose it, and considers the term "sub-species" the "more appropriate" (Darwin 280).

As Dennis Porter has written of Darwin's Fuegian observations, "Darwin's capacity for intraspecies fellow feeling reaches its limit; there is, so to speak, a shock of nonrecognition" (161). In Louisa Meredith's first- and secondhand observations of Aboriginal people at Bathurst in New South Wales in her text of 1844, she points out the existence of a "great diversity of dialects," as well as music "with much tune and variety" (*Notes and Sketches* 99), but these observations could simply imply a greater body of preexisting knowledge about the Aborigines than about the Fuegians. Nevertheless while her general tone does incline toward a "single perception of physical type," an inclination that is to strengthen with the passage of years and the concurrent prevailing currency of anthropological attitudes, there is no shock of nonrecognition; indeed, she displays distinct sympathy for the women especially, who occupy "as miserable and debased a position . . . as it is possible for human beings to do" (*Notes and Sketches* 93).

Simon Gikandi, in discussing Charles Kingsley's *At Last: A Christmas in the West Indies* (1885), shows how Kingsley "operates from the natural historian's premise that the study of animals and plants, rather than the observation of human life, provides us with deeper meanings about the order of nature and things" (Gikandi 57), an activity Gikandi describes as "showcasing the colonial space" (57). Meredith's earliest English publications, those botanical studies of English domestic and wild flowers, elaborately illustrated with colored plates, could be described as showcasing productions from the domestic space. From Tasmania she later produced similar elaborate, highly ornate books of illustrated studies of Australian flowers, insects, and animals, showcasing the colonial space as a new treasure-house of rare and beautiful flora and fauna.

In 1891 Macmillan published Meredith's *Last Series: Bush Friends in Tasmania, Native Flowers, Fruits and Insects, Drawn from Nature with Prose Descriptions and Illustrations in Verse.* She was seventy-nine years old, and it was to be her last publication. Among the plants she discusses in this work is the "Currijong," a plant like the famed New Zealand flax, and she recalls that she sent a sample

> home to the great exhibition of 1862 some of it in its raw state, and another specimen, which had been roughly twisted into cords by one of our shepherds, who used it for making his "snares" for opossums. It was a pleasant surprise to receive the grand bronze medal for my little exhibit, as being "a new material for paper." (Meredith, *Bush Friends* 10)

This is economic botany taken to extremes. As Brockway explains, every "new plant was being scrutinized for its use as food, fiber, timber, dye, or medicine" (74). I read a vague sense of retrospective surprise on Meredith's part that her shepherd's rope could have been construed as a form of paper, although she is happy enough to have a medal anyway, and the incident confirms the way in which the colonial space was being quite mindlessly ransacked for possible profitable yield.

Plate VIII in *Bush Friends* is titled "A Cool Debate" (figure 6.2). Four frogs are sitting on lily pads forming an apparent round table. Beneath this title Meredith has provided the Latin names for the flowers and frogs in the painting. Her practice recalls that of Thomas Bell in *Zoology of the Voyage of the Beagle* and shows how unremarked the use of scientific labels has become in the intervening decades between 1844 and 1891. In her preface, indeed, Meredith now thanks distinguished men

Figure 6.2. *A Cool Debate*. Lithograph of the Australian gold frog and the native buckbean. Plate 8, *Last Series. Bush Friends in Tasmania: Native Flowers, Fruits, and Insects, Drawn from Nature with Prose Descriptions and Illustrations in Verse* (1891). University of Western Australia Library.

of science: "Sir Joseph D. Hooker, C.B., K.C.S.I., F.R.S., F.L.S. . . . for his kindness in looking over my proof-sheets, to prevent botanical errors," and Professor J. O. Westwood, M.A., for providing "the proper names of the insect figures in my pages" (v). Their imprimatur authorizes her work as scientific, and she need make no coy prefatory disclaimer of the kind she always resorted to in the 1830s and 1840s. She does however call herself in passing a "comparatively ignorant enthusiast" (*Bush Friends* 13), in comparison, that is, to her late friend Dr. Joseph Milligan, F.R.S., F.G.S.

"A Cool Debate" includes the color portrait, which is accompanied by a prose description, and a poem. The prose section of "A Cool Debate" refers to "my frog parliament" and "my grave and portly friends, in their shining Court suits," and in yet more queenly style Meredith writes, "my magnificent favourites sat for me" (*Bush Friends* 39–40).[7] The use of the possessive is absent from her 1844 text, *Notes and Sketches of New South Wales*, and its reiteration here suggests Meredith's successful shift from traveler to powerful colonizer. The prose section also includes occasional biological description of the way the eyes of the frogs function, passages in which the discourse shifts between the scientific and the comic:

> Their eyes are large and bright, with a change and variety of expression that is queerly intelligent, and their winking was to me very droll. As the eye closes, the whole dome-like excrescence which contains it drops down to the smooth level of the head, as if suddenly engulfed by a cerebral trap-door, and then—literally in the twinkling of an eye—it pops up again, and looks gravely round . . . the effect was grotesque and startling. (*Bush Friends* 40)

The general appearance of the frogs, however, is detailed in the 1891 text using identical words from the 1844 text, including "gorgeous," "embroidery," "shaded," "velvet," and "delicate." But the overall tone has become much less chivalric, the general term "array" has now been replaced with the more specific "coats," and Meredith adds, "the waistcoat portion of their garb I may describe as white kid, mottled with delicate creamy shades, and marbled with thin gold leaf" (*Bush Friends* 41), a description that uncannily suggests a Regency buck. Even tadpoles warrant a mention: "Some human exquisites delight in adorning themselves with jewels; the green tadpole is himself an embodied jewel" (*Bush Friends* 41). By degrees the terms Meredith is using to describe frogs in 1891 are the same as those she uses to write of indigenous people in 1844; specifically, the lexical units "beaux" and "exquisites" recur. I call these "units" because the words carry particular historical baggage. The shift is most apparent in the poem she produces for "A Cool Debate," which she titles "Water Beaux":

> Ah! do you laugh, because I say these frogs
> Are Water Beaux? In sooth, I do not know
> That mortal exquisite, who pranks himself
> In comparable bravery of garb.
> Broadcloth and velvet are prosaic, dull,
> Nowhere in competition! (*Bush Friends* 43)

The shift in lexis signals a particular absence from the colonial space—that is, from the Tasmanian colonial consciousness—in 1891. As far as these colonizers are concerned (while not actually the case), the colonized people, the original inhabitants of Tasmania, have disappeared. This consciousness, combined with the practice of ethnographic showcasing in which animals and native people were lumped together in European zoos, circuses, and exhibitions, reveals not only, in Ghassan Hage's terms, "the power of the nation that exhibited them" (130), but also the reason why the same lexis can be applied to animals and native peoples indiscriminately. Hage continues: "The 'exotic peoples' exhibitions at the colonial world fairs operated much like the zoo in that those exhibited were seen as both part of subjugated 'nature' (natural people) and as colonial 'artefacts' " (131). No one was more familiar with showcasing the products of the colonial space than Meredith, in terms of both publication and exhibition. She had sent botanical paintings to the Great Exhibition of London in 1862; she was the only woman to win a silver medal for art combined with literature at the Inter-Colonial Exhibition held in Victoria in 1866 and 1867; and she received awards from exhibitions in Sydney (1870), Melbourne (1881), and Calcutta (1884) (Rae-Ellis 209).

Stocking observes that in the middle years of the nineteenth century, broader social and colonial contexts conditioned Anglo-Saxon racial attitudes. Events like the Indian Mutiny sharpened racial resentments; "nigger" as an epithet came into more general use, and the American civil war over slavery polarized perceptions of race (63). Thus what emerged was a "racialism of a harsher, hereditarian sort," which denied that all people had a common origin, and established non-Europeans as "of permanently inferior capacity, impervious to social influences in the present" (107).

The power of this hardening attitude and its continuance are evidenced by Louisa Meredith's *Bush Friends* (1891), supposedly about flora and fauna, as I have shown in describing "A Cool Debate." This work abruptly fissures at one point, however, to reveal her angry sensitivity about what must have been a current debate at the time, the accusation that genocide had taken place in Tasmania. In justification of the rightness of past colonial actions in Tasmania, actions of which she has only secondhand knowledge, Meredith records a horrific story of an attack by Aboriginal people on white settlers, including a painstaking detailing of injuries and mutilations based solely on hearsay. She then declares that she has never heard of comparable stories of "the death or injury of one of the Aborigines," attributing tales of such butchery to "sensational writers" (*Bush Friends* 52). She offers her explanation for the apparent absence of indigenous people, in a discourse that is a classic of its kind:

> the remaining natives, always few in number, had been peaceably removed to Flinders Island, in Bass Straits, under careful and kindly supervision, with supplies of food and clothing, and with abundance of wild animals and fish to hunt and capture. (53)

She describes colonial existence before this enforced removal as a "reign of terror" in which occurred "cunning, cowardly murders of solitary shepherds . . . helpless old people and little children" (53).

Meredith summarizes in popularist pseudo-ethnological terms her version of Tasmania's indigenous people as "surely the very lowest creatures in human form. Their countenances, as shewn by the excellent photographs of the last four, made at my suggestion for the Melbourne Exhibition many years ago, bore a curiously close resemblance to pug-dogs" (*Bush Friends* 53). Aware that this subject is not in keeping with the project of her book as a whole, she nevertheless asserts the truth of her claims, writing, "my *true* "Old Story" needed some explanatory remark" (the emphasis is Meredith's), and adding that she could not resist offering

> such testimony as my own experience affords, in a negative, circumstantial way in defence of those brave-hearted early Colonists whose acts and characters have been so hardly, and, as I must believe, so unjustly attacked. They were not a band of lawless, merciless, desperadoes, but, for the most part, honest, peaceful, God-fearing country gentlemen and yeomen. (53)

She rounds off this dubious "testimony," which she has herself condemned with the terms "negative" and "circumstantial," with another poem very different in tone from her frog poem "Water Beaux," which she titles "An Old Story of 1834." The epigraph to this poem, "Pity 'tis, 'tis true" (55), reasserts her claim to the truth of the story she tells, and the poem itself repeats the prose account of savage attacks on white settlers. The poetry is of the regular ballad style, with five-line stanzas and a simple rhyme scheme, very different from the complex syntax and blank verse of the earlier poem on frogs. This ballad style emphasizes her portrayal of the early colonists as simple folk and reinforces a pathetic sentimentality. The chief persona is an old shepherd who tells the "true" story:

> Another morning—and again
> The shepherd, whistling, clomb
> The rugged pathway, whence his eye
> Glanced where the brook ran merrily
> Past Will's snug cottage home.
>
> He looked across the sunny knoll,
> Where erst the logs were seen:
> He started! Four stumps, black and crook'd,
> Stretched boughs all stunted, gnarled and hook'd,
> Athwart the herbage green.
>
> He stared—and doubted—rubbed his eyes—
> Turned—gazed again—but still

> Those ugly shapes, black, weird and grim,
> Stood rigidly confronting him
> Upon the rounded hill. (57–58)

The "stumps" are of course native people on the watch, as the grim semantics of the description clearly indicate, people who are also designated in this poem as "those black devils" and "cursed imps of the night" (59, 60). The lexical shift from "elvish" in 1844 to the more negative "imp" in 1891 pronounces the distance Meredith has traveled in hardening racist attitudes. A concluding stanza suggests that had any aboriginal people survived to testify, they would themselves reject the accusations of butchery made by sensation writers:

> The saxon foot, with onward step,
> Another race treads out,
> Which scarce a vestige leaves, to tell
> It ever was—or to dispel
> The Future's curious doubt. (62)

The reviewer of the first number of the *Tasmanian Journal of Natural Science* in 1842 enthused about the way in which colonization would widen and cultivate the empire of science. When Louisa Anne Meredith writes about what she knows and observes at first hand, her informed botanizing and delighted detailing of insect life lend her prose an integrity reflected in Elizabeth Rigby's enthusiastic review: "we can only sum up her tarantulas, her scorpions, her ants, spiders, crabs, and grubs, and all kinds of other nasty things, with the unqualified assertion that nobody ever made them so nice before" (107). However, when she attempts to extend the boundaries of her scientific empire beyond "the poetry of frogs," to write on the indigenous people of Tasmania with whom she has had no firsthand experience and little if any contact, she produces only prejudicial and distorted versions of colonizing history. She abandons all pretense to objective discourse for a racist rhetoric housed in the most inappropriate form imaginable, that of poetry.

Louisa Anne Meredith (née Twamley) (1812–1895): Major Works

England (as Louisa Anne Twamley)
The Romance of Nature; or, The Flower Seasons Illustrated (1836)
Flora's Gems; or, The Treasure of the Parterre (1837)
Our Wild Flowers Familiarly Described and Illustrated (1839)
An Autumn Ramble by the Wye (1839)

Australia (as Mrs. Meredith)

Notes and Sketches of New South Wales during a Residence in That Colony from 1839 to 1844 (1844)
My Home in Tasmania; or, Nine Years in Australia (1852)
Some of My Bush Friends in Tasmania (1860)
Over the Straits: A Visit to Victoria (1861)
Our Island Home: A Tasmanian Sketch Book (1879)
Tasmanian Friends and Foes, Feathered, Furred, and Finned (1880)
Last Series. Bush Friends in Tasmania (1891)

Notes

1. Ann Shteir's recent publication *Cultivating Women* extends and elaborates on the ideas in this earlier essay.
2. In an earlier article, " 'Woman's Testimony,' " I discuss how this review helps to formulate a female imperialist discourse. In this discussion my focus is on Rigby's construction of the archetypal English Woman.
3. I recognise that neither "indigenous" nor "Aboriginal" are satisfactory terms but in the absence of any better interchange the two. I also use the terms "local people" and "native people."
4. The Brothers Grimm published "The Frog Prince" in 1823.
5. Rowland's Macassar oil was "grandiloquently advertised in the early part of the 19th century, and represented by the makers (Rowland and Son) to consist of ingredients obtained from Macassar" (*OED*). It is this that inspired Lord Byron, *Don Juan*, canto 1:17, who writes of Don Juan's mother: Oh! she was perfect past all parallel— / Of any modern female saint's comparison; / So far above the cunning powers of hell, / Her guardian angel had given up his garrison; / Even her minutest motions went as well / As those of the best timepiece made by Harrison: / In virtues nothing earthly could surpass her, / Save thine "incomparable oil," Macasser!
6. The friend of Byron and Lady Blessington and "the beau ideal of manly dignity and grace" according to his contemporaries (*Dictionary of National Biography*).
7. As Meredith's biographer Vivienne Rae-Ellis remarks, photographs of Louisa Anne Meredith in old age show a striking resemblance to Queen Victoria (Rae-Ellis 222).

Works Cited

"Aborigines of New South Wales," *Hogg's Weekly Instructor* 1 (1845): 63.

Bell, Thomas. "Part V: Reptiles." *Zoology of the Voyage of the Beagle*. Parts 1–5. Facsimile Reprint Wellington, NZ, 1980, from London edition of Smith Elder, 1843.

Brockway, Lucile H. *Science and Colonial Expansion: The Role of the British Royal Botanic Gardens*. New York: Academic, 1979.

Chambers, Robert. *Vestiges of the Natural History of Creation* (1844). Leicester: Leicester UP, 1969.
Chorley, Henry. Review of *Notes and Sketches of New South Wales, Athenaeum* (January 1845): 66–67.
Darwin, Charles. *The Descent of Man, and Selection in Relation to Sex* (1871). London: John Murray, 1906.
"Frogs, Frog-Tribe." *Penny Cyclopaedia of the Society for the Diffusion of Useful Knowledge*. Vol. 10. London: Charles Knight, 1838.
Gikandi, Simon. "Englishness, Travel and Theory: Writing the West Indies in the Nineteenth Century." *Nineteenth-Century Contexts: Colonialisms* 18 (1994): 49–70.
Hage, Ghassan. "Republicanism, Multiculturalism, Zoology." *Communal/Plural* 2 (1993): 113–37.
Johnston, Judith. " 'Woman's Testimony': Imperialist Discourse in the Professional Colonial Travel Writing of Louisa Anne Meredith and Catharine Parr Traill." *Australian and New Zealand Studies in Canada* 11 (1994): 34–55.
Meredith, Louisa Anne. *Last Series. Bush Friends in Tasmania: Native Flowers, Fruits, and Insects. Drawn from Nature with Prose Descriptions and Illustrations in Verse*. London: Macmillan, 1891.
Meredith, Louisa Anne. *Notes and Sketches of New South Wales during a Residence in That Colony from 1839 to 1844*. Sydney: Ure Smith, 1973.
Meredith, Louisa Anne [Twamley]. *The Romance of Nature; or, The Flower Seasons Illustrated*. London: Tilt, 1839.
Mills, Sara. *Discourses of Difference. An Analysis of Women's Travel Writing and Colonialism*. London: Routledge, 1991.
Porter, Dennis. *Haunted Journeys. Desire and Transgression in European Travel Writing*. Princeton: Princeton UP, 1991.
Rae-Ellis, Vivienne. *Louisa Anne Meredith: A Tigress in Exile*. Tasmania: Blubberhead P, 1979.
Rigby, Elizabeth. "Lady Travellers." *Quarterly Review* 76 (1845): 98–137.
Shteir, Ann B. *Cultivating Women, Cultivating Science: Flora's Daughters and Botany in England, 1760 to 1860* (Baltimore: Johns Hopkins UP, 1996).
Shteir, Ann B. "Linnaeus's Daughters: Women and British Botany." In *Women and the Structure of Society*, ed. Barbara J. Harris and JoAnn K. McNamara. Durham, NC: Duke UP, 1984.
Stocking, George W., Jr. *Victorian Anthropology*. London: Macmillan, 1989.
"Tasmanian Journal." *Monthly Review* 157 (1842): 520–33.
Traill, Catharine Parr. *The Backwoods of Canada: Being Letters from the Wife of an Emigrant Officer, Illustrative of the Domestic Economy of British North America*. London: Charles Knight, 1836.

7

"Through Books to Nature"

Anna Botsford Comstock and the Nature Study Movement

Pamela M. Henson

"Anna, heaven may be a happier place than the earth, but it cannot be more beautiful" (Anna Botsford Comstock, *Comstocks of Cornell* 57), Phoebe Botsford remarked to her daughter as they watched the sun close the day with a magnificent display of color and light. Those words and the love of nature they conveyed had a lifelong influence on Anna Botsford Comstock (1854–1930), bringing her daily joy in the world around her, a marriage to a leading entomologist that was centered on the study of nature, and a career that had a significant influence on popular attitudes toward the environment. Building on a reverence for the natural world instilled by her mother, Anna Comstock became a noted nature artist, a leader in the field of nature study, and one of the most influential popular science writers in the United States. In the process she combined her artistic talents, love of literature, knowledge of science, and interest in education to create a new approach to the study of nature that combined accurate knowledge with an aesthetic appreciation of and personal relationship to the natural world.

Anna Comstock entered science through the "back door," as many female relatives of scientists did, and she always worked on "the peripheries" of science in art, popularization, and children's education. Her choice of this career path seems to have been determined by a combination of available opportunities for women and her own stated

Figure 7.1. Anna Botsford Comstock. Courtesy of the Charles P. Alexander Papers, Smithsonian Institution Archives.

preferences for women's work and a personal relationship to nature. In developing her skills as an illustrator, she assumed a socially acceptable role for women in science. As Kohlstedt and Shteir have demonstrated, a number of eighteenth- and nineteenth-century women aided their husbands' and fathers' careers by preparing their illustrations (Kohlstedt, "Periphery" 86–96; Shteir, "Linnaeus's Daughters" 70–73). Working in a domestic setting, using talents deemed feminine, these women used their artistic skills to produce the images required in their male relatives' publications. The women in the botanical Hooker and Turner families in England contributed to the "family firm" in this way (Shteir, "Botany" 34–36). In the United States, Lucy Say illustrated her husband's *Conchology,* and Sarah Aiken Hall prepared plates for her husband John's work on the New York State Geological Survey (Kohlstedt, "Parlors" 425–45). Such artistic work was deemed a proper sphere for ladies because it was done privately and did not intrude into traditionally male domains of scientific research.

Science writing by women for children had come into vogue in late-eighteenth-century England, with authors like Lady Elleanor Fenn, Priscilla Wakefield, Laetitia Ford, and Maria Jacson, who used accounts of animal and plant life to teach children morals and virtues, as well as science. *An Introduction to the Natural History and Classification of Insects* written in 1816 by the Quaker Priscilla Wakefield was set as an exchange of letters between two sisters and was designed to improve the minds and characters of both boys and girls. By mid-nineteenth century, nature writing had become an important part of the popular press in the United States in such magazines as *Harper's Weekly, St. Nicholas,* and *Youth's Companion,* opening doors for women (Whalen and Tobin 195–203). In the 1870s, Mary Treat, a naturalist in New Jersey, began publishing her studies of the flora and fauna of the New Jersey Pine Barrens in professional journals. But Treat's greatest successes, financially, were her popular science writings for children and adults, such as *Home Studies in Nature* and *The Butterfly Hunters,* and these may have provided a more satisfying model for Anna Comstock to adopt (Bonta 42–48; Mallis 80).[1]

Comstock chose to stay on the peripheries of science for both positive and negative reasons. Opportunities for women were genuinely limited, although not entirely closed. In addition, scientific research required adopting an "objective" approach to nature that she does not appear to have found very satisfying. As a very creative individual, Comstock found it more comfortable to incorporate her aesthetic appreciation of nature into scientific interpretation for children and a popular audience. This appreciation was part of her overall sense of subjective connectedness to the world around her. Anna Comstock experienced the natural world in emotional terms and felt a sense of personal relationship and responsibility to the living things around her. Keller and others have argued that objective science became identified with masculine characteristics, while aesthetic and emotional relations were regarded as feminine (Keller 72–94). While following her own preferences and talents, Comstock certainly did choose appropriately gendered roles.

Illustration, popular writing, and teaching afforded Comstock opportunities to synthesize her visual and literary aesthetic, love of nature, knowledge of animal and plant life, and ideas about education. Her writings and visuals take four distinct forms: (1) work for her husband's scientific publications; (2) children's writings; (3) nature study lessons; and (4) popular adult science writings. She began by illustrating and then writing for her husband's publications, then moved into writing about nature for children. The combination of her experience with her husband and her children's writing led to her involvement in the nature study movement and, later, to popular science writing for adults. Eventually her professional demands forced her to give up work on her husband's publications, but she continued to write for all three of her own audiences. During this long career, she moved from standard prose and illustration to a distinctive drawing and writing style that appealed to a broad popular audience.

The path that Comstock followed to a career as a popular science writer was a winding one, built upon and altered by important influences and interests in her life. She set out many features of this story in her autobiography, *The Comstocks of Cornell*. Raised on a farm in upstate New York at mid-century, she accompanied her mother on nature walks through the nearby fields and woods, learning the names of wildflowers and the constellations of the night sky. Her Quaker mother studied to improve herself and her family, lulling her daughter to sleep with poetry. Valuing education, the Botsfords sent Anna to Cornell University in 1874, shortly after it began to admit women, making her one of the first generation of women in the United States to secure a college education. Even as a child, she delighted in literature and poetry, especially the romantics, transcendentalists, and nature poets. When at age fourteen she earned the magnificent sum of eighteen dollars, she promptly spent it on volumes of Shakespeare, Moore, Burns, Byron, Scott, Tennyson, Longfellow, and Dickens.

She pursued her love of the written word at Cornell, influenced especially by Professor Hiram Corson, but after her second year, she left college and returned to her parents' home, leaving no explanation for this decision. The path of her life had, however, already been set. In addition to her literature courses, she had studied zoology with John Henry Comstock, a young professor of entomology. They had also spent many an afternoon on nature walks in the woods surrounding campus, collecting plants and insects. Their mutual interests in the natural world and delight in one another led to marriage and a shared career in science at Cornell.

Even before their betrothal, John Henry Comstock sent Anna art supplies so she could learn to draw insects for him. After their marriage in 1878, she assumed the role of helpmate, washing bottles in the lab, accompanying him on field trips, preparing research notes and class handouts. When the hoped-for children did not appear, she spent more and more time assisting her husband and taught herself wood engraving. When John Henry served as Chief Entomologist at the U.S. Department of Agriculture in 1880–81, Anna's work assisting him was so highly regarded that the commissioner of agriculture placed her on the payroll. This proved a pivotal point in her life as she began to take her own work more seriously. After the Comstocks returned to Ithaca, she completed her bachelor's degree with a thesis in entomology, but she did not pursue a career as a research scientist. Instead she developed her skills as an artist, studying in New York City with the noted wood engraver John P. Davis. She began lecturing and writing on science for a popular audience, interpreting scientific knowledge rather than adding to it. She also took advantage of opportunities to work in new areas of children's and adult education.

The door to scientific research was not entirely closed to Anna Comstock, however, since she did have a role model for a woman scientist in her close friend Susanna Phelps Gage, a noted embryologist. Married to John Henry Comstock's best friend, Simon H. Gage, Susanna pursued a distinguished research career but never received a paid position (*Comstocks of Cornell* 82–180, 249; Conable 85, 89). Perhaps the path followed by such earlier women nature writers as Wakefield and Treat appeared not only more promising to Anna Comstock but also more lucrative.

Anna Comstock did articulate very clear views about appropriate roles for women. She always seemed more comfortable with traditional women's roles, and she wrote that women should focus their energies on children's education. Although a pioneer in many ways, she did not actively support universal suffrage for women, arguing that women should address local issues, leaving national politics to the men. Her caution may have been the result of rebuffs, as her infrequent moves out of these safer roles met resistance. In 1898 she had the singular honor of being appointed the first woman professor at Cornell, but the board of trustees promptly revoked the title. Adamant that women could not enter the professorial fraternity, Cornell did not appoint any women professors until 1911 and then only in home economics. Comstock regained her professorial title only in 1913, after working for many years as a lecturer (Conable 127, 130).

Anna Comstock's most intriguing writing about the role of women can be found in her one novel, *Confessions to a Heathen Idol* (1906). A satire of her experiences in academic life, first published under a pseudonym, *Confessions* traced the life and loves of a young widow of a college professor. Each evening she "confesses" to a teakwood idol on her desk as she tries to sort out the complexities of her life. As she and her mother-in-law contemplate remarriage, they debate the comparative advantages of married versus single life. They concur that considerable advantages attend widowhood, which combined the social status of a married woman and the freedom of a single one. In the end, both choose true love and remarriage, but with a clear understanding of the compromises and sacrifices women must make.

In her real life, Anna Comstock first chose marriage and then found ways within it to utilize talents and energies that complemented her husband's work. In the 1880s, she began to write essays about the insect world for popular children's magazines such as *St. Nicholas*. She also accompanied her husband to Farmer's Institutes, where he lectured on insects and agriculture. Soon she was drafted into lecturing to farmers' wives on the values of rural life. Children's writing and agricultural extension work were both fields open to women at that time; thus, she was able to take on more independent work without stepping outside the bounds of gender-appropriate roles. But both Comstocks had come to regard her work as an illustrator as an integral part of his scientific work. Indeed, when her insect illustrations were published in his 1888 textbook, *An Introduction to Entomology,* Anna Comstock was given credit for her "original drawings and engravings" on the title page—a recognition not often afforded the wives, daughters, and sisters of scientists.[2]

In her popular children's writings, Comstock used her considerable knowledge of the natural world to weave tales about animal and plant life in lively and charmingly illustrated dramas. In these writings, she anthropomorphized animal life, giving our fellow creatures very human emotions, physical characteristics, and motivations. Although influenced by earlier nature writers, Comstock developed her own narrative style for her children's essays. Wakefield and other women writers had used epistolary or conversational formats among imaginary family members to engage children in learning science. Comstock preferred to weave tales of nature as drama unfolding around the reader at every turn and in every tree. Her essay "The Story We Love Best" in *The Ways of the Six-Footed* (1903) contains all the essential elements that characterize Anna Comstock's popular science writings. The carpenter bee is a familiar creature, found in sumac, elder, or raspberries near any rural home, but she noted, as familiar and uninteresting as these branches may seem, they "have been the scenes of great toil, brave deeds, faithful devotion, and also, alas, of treachery and tragedy" (108).

Comstock therefore turned the life history of the carpenter bee into nature's drama, as a mother bee prepares a nest for her eggs, provisions and protects them, loses some to predators, takes her offspring on their maiden flight, and protects them over their winter rest. Unlike her husband's impersonal description of the same story in another popular book (*Insect Life*), Comstock's story chronicled the activities of the mother bee in very human terms. "Thus, you see, this matron has her family in an apartment house, each child occupying an entire flat" (*Ways of the Six-Footed* 111). Describing the bee's relentless efforts to care for her young, Comstock declared, "to us she is more than a mere creature of instinct; she is a bee of character" (116) who yields "the last energies of her ebbing life in unselfish devotion to her home and family" (117). Although based on careful observations and knowledge of Ceratina biology, the story anthropomorphizes the tiny creatures, endowing them with human personality characteristics as they play out the drama of life and death in a sumac branch, thereby creating a sense of empathy and interest on the part of her young reader. This essay also taught readers the gender characteristics that Comstock believed were appropriate to motherhood: fidelity, perseverance, industry, patience, pluck, faithfulness, unselfishness, and devotion to home and family.

In describing the female mud dauber wasp, Comstock told her reader that "they have waists so slender and aristocratic that they would make figures of a Parisian fashion-plate seem coarse in comparison" (*Ways of the Six-Footed* 96). She speculated that this insect's fidgety nature might be due to "the burden of family cares; they all fall upon her, as her spouse is a short-lived, indolent fellow, who never dreams of lending her a helping mandible in the construction of their home" (97). Thus, she noted, the issue of "women's rights" is settled in the affirmative in the insect world. These anthropomorphic characterizations did appeal to a juvenile audience, but they also weakened Comstock's claims to scientific accuracy and opened her work to criticism as overly sentimental.

Comstock's illustrations often echoed her sense of whimsy and drama in the animal world. Her essays were charmingly illustrated with anthropomorphized creatures, such as a musical cicada in *Ways of the Six-Footed,* that captured visually the tone of her writing (140; figure 7.2.) After his seventeen years of isolation the cicada emerges, and "So happy is he that he feels as if he must burst if he does not find some adequate means for expressing his happiness in the beautiful world of sunshine. Then suddenly he finds in himself the means of

X

HERMIT AND TROUBADOUR

A JUNE STORY FOR JUNIOR NATURALISTS

N far Thibet exists a class of Buddhist monks who are hermits and who dwell in caves. I was told about these strange people by a senior naturalist, who has spent his life going around the world and finding the countries upon it as easily as you junior naturalists find the same countries on the globe in the schoolroom. A real naturalist is never contented with maps of places and pictures of things, but always desires to see the places and things themselves.

The senior naturalist told me that he found Thibet a dreary land inhabited by queer people; and the hermit monks were the queerest of all. Each one dwelt in his solitary cave, ate very little, and worked not at all, but spent his time in thought. Could we

Figure 7.2. "Hermit and Troubadour." From *Ways of the Six-Footed.*

expression and bursts into song" (146). If the cicada could talk to a junior naturalist,

> He would perhaps tell you also that he had his eye on a certain graceful maiden perched on the leaf between him and the sun; but she, on the other hand, seemed to give about equal attention to him and three other drummers situated nearby. Excited by the competition and by her indifference he rattled his drum faster and faster until he arose to heights of cicada melody and harmony that left his rivals far behind. Then the lady of his choice listened spellbound and pronounced him the greatest of all musicians, and thus he won his bride. (147–48)

As Anna Comstock developed her skills as a writer, her distinctive voice also began to appear in the Comstocks' scientific textbooks. In the *Manual for the Study of Insects* (1895), for example, she added life history accounts to her husband's anatomical descriptions and is listed as coauthor, not just illustrator. Her voice can be clearly distinguished from his, as in the following paragraphs on the bumblebee. Her husband wrote, "In the Apidae we find that the lower lip has been highly specialized for the procuring of nectar from deep flowers. Here the glossa is slender and greatly elongate, being longer than the mentum . . . ; the basal segments of the labial palpi are also elongate (666–67). While John Henry used objective and technical language, Anna Comstock's description of how these bees live invoked a personal relationship to the animal, revealing her own delight in their habits: "The clumsy rover, the bumblebee, is an old friend of us all. As children we caught her off thistle-blossoms and imprisoned her in emptied milkweek pods, and bade her sing for us. . . . And she has deserved all the attention and affection bestowed upon her, because she is usually good-natured and companionable. She is a happy-go-lucky insect, and takes life as it comes without any of the severe disciplining and exact methods of her cousin, the honey-bee" (672–73).

This tone and style of writing were not at all typical of the scientific texts of the day, such as Alpheus Spring Packard's *Guide to the Study of Insects* (1869), a competing text written in technical prose. Packard's volume was a mixture of his own observations and extensive quotations from earlier entomologists. Even in the life history sections, the tone of his description of the bumblebee is similar to John Henry Comstock's: "After the first brood, composed of workers has come forth, the queen bee devotes her time principally to her duties at home, the workers supplying the colony with honey and pollen. As the queen continues prolific, more workers are added, and the nest is rapidly enlarged" (131). Packard occasionally offered value judgments on an insect's activities, but in a far more subdued tone than Anna Comstock's. Of the Ceratina bee—the same bee that Anna Comstock found so full of character in her children's essay—he wrote, "This bee is oblong in form, with tridentate mandibles, and a short labrum. The maxillary palpi are six-jointed, and the labial palpi are two-jointed" (134). But he continued, "In none of the wild bees are the cells constructed with more nicety than those of our little Ceratina. She bores out with her jaws a long deep well just the size of her body, and then stretches a thin delicate cloth of silk, drawn tight as a drumhead, across each of her chambers, which she then fills with a mixture of pollen and honey" (135).

In contrast, Anna Comstock's distinctive personal voice was a significant departure from the usual style of science textbook writing of that era. Although found in their joint publications, this style of writing was not found in her husband's earlier or later writings, even those, such as *Insect Life*, that were aimed at a popular audience. Anna did not, then, change John Henry's writing style; the tone of writing in his volumes changed only when she was a coauthor. Although Comstock's writing style proved popular in the late nineteenth century, anthropomorphism and sentimentality ultimately brought the nature study movement under attack from male scientists in the 1930s. The new generation of scientists attempted to limit the right to speak about science to a narrow community of professionals, predominately male and trained in the "objective" science tradition. They ridiculed the nature study movement as unscientific, sentimental, and inaccurate, as they gained control over the science education establishment.

Anna Comstock's status as a coauthor was unusual for scientific wives of that era, but it reflected how seriously both Comstocks took her contributions. And during the 1890s Anna Comstock assumed a truly professional role on several fronts, leading to her appointment in 1898 as the first woman professor at Cornell. She developed her skills as an artist, producing scenes of the natural world around her, from brooks, to forests, to moths against the night sky. Her work was displayed at exhibitions worldwide and won her international acclaim as a nature artist. Both Comstocks were visual thinkers who emphasized observation over rote memorization in learning. Thus, they regarded drawings that illustrated the text as crucial to their publications. As visual learners they were both drawn to the Pestalozzian method of education, which emphasized direct experience of objects rather than book learning and memorization. Books and drawings were to be used as tools to guide experiential learning. The Comstocks set up their own publishing house, Comstock Publishing Associates, to ensure the quality and quantity of illustrations they wanted in their books. Their motto for the new venture, "Through books to nature," was a modification of Louis Agassiz's dictum, "Study nature, not books." They viewed books and drawings as essential tools to introduce teachers, students, and adults with leisure time to the beauties and complexities of nature. Science writing as an important component of teaching science, but it needed to be complemented by direct interaction with nature (Henson, "Evolution and Taxonomy" 267–75; Palmer 7–68).

Anna Comstock was given an opportunity to place these ideas into action in 1895 when she was asked by the Committee for the Promotion of Agriculture in New York State to help develop a school curriculum on "Nature-study" and rural life that would encourage farm youth to stay at home rather than migrate to overcrowded cities. By teaching rural children to observe and appreciate the world around them, nature study would tie rural youth to the land and discourage migration to urban slums. Using her knowledge of science, agricultural extension experience, and talents as a children's writer, Comstock joined forces with Liberty Hyde Bailey to create a center for the nature study movement at the School of Agriculture at Cornell. Under Comstock's tutelage, a number of women, including Flora Rose, Alice McCloskey, and Martha Van Rensselaer, pursued professional careers in this new field (Palmer 3, 39–43). Comstock's writings in the *Cornell Nature-Study Leaflets, Cornell Nature-Study Bulletin,* and *Nature Study Review* reached a national audience of elementary and secondary school teachers, urban and rural. Her *Handbook of Nature Study,* published in 1911, became the "Nature-study Bible," translated into eight languages and printed in North and South America, Asia, and Europe. Indeed, it remains in print today.

The nature study movement proved an ideal vehicle for Comstock because its goal was not to train future scientists but to teach children to be familiar with and enjoy the natural world around them. As Liberty Hyde Bailey wrote in *The Nature-Study Idea* (1903), nature could be studied with either of two objectives: "to discover new truth for the purpose of increasing the sum of human knowledge" or "to put the pupil in a sympathetic attitude toward nature for the purpose of increasing his joy of living" (4–5). The goal of nature study was the latter, to change attitudes, not teach facts. The nature study student need not become a scientist or ever make an original discovery, but she should develop a sympathy for nature that enriches daily life. With its emphasis on learning to appreciate nature, rather than dissect or control it, nature study provided the ideal outlet for Comstock's intellectual and creative energies.

Her nature study lessons were more serious in tone than her children's writings and were designed as study guides. The lessons opened with background information on a topic, such as how birds fly, followed by instructions for observations and questions to guide analysis of the topic. Although her focus was still on the domestic—birds, flowers, and rocks that could be found in the reader's own yard—Comstock moved beyond insects to encompass all of the animal, plant, and mineral world. In each section, she pointed out the many benefits offered by the animal or plant—recycling garbage, controlling other pests, and maintaining the balance of nature. With the advent of halftone photography, she was more likely to illustrate these lessons with photographs than with new engravings, but her sense of the aesthetic still came through in the nature poetry that dotted these essays. Well illustrated, the nature study lessons went through several editions, first as the *Cornell Nature-Study Leaflets*, then as a series in the *Chautauquan*, and finally as her magnum opus, the *Handbook of Nature Study*.

The 1890s saw the advent of the Chautauqua movement, which afforded Comstock the opportunity to reach an even broader audience. She crisscrossed the country on this lecture circuit and published regularly in the *Chautauquan*. This popular adult education movement gave her national visibility as an interpreter of science. As the market for popular science grew at the turn of the century, she was in great demand as a writer and editor for magazines like *Country Life in America* and for such publishers as D. Appleton's, Ginn and Company, and Doubleday, Page, and Company. She became so busy that she stopped producing her husband's illustrations, and the only works that they coauthored were in her areas of nature study and popular science writing. In the years from 1895–1930, Anna Comstock became an internationally known lecturer, writer, teacher, and editor; indeed by 1923 Anna Comstock was so well-known that she was voted one of the twelve most respected women in America in a League of Women Voters poll.[3]

Through her adult popular writings, Anna Comstock reached an even broader audience with her message of love for the natural world. Despite her early emphasis on rural life, she found a ready market in urban and suburban homes where experience of nature was limited, but families had new leisure time to spend on outdoor pursuits. The automobile allowed families to travel to new places and enjoy the out-of-doors away from their homes. Anna Comstock modified her writing to focus on the recreational aspects of nature study, as she taught her readers how to observe nature from a train or car window while traveling and described wildlife in more exotic locales, from the deserts of the Southwest to Hawaii. In these essays, she moved from a domestic setting to guide Americans as they took to the roads with their new leisure time. In "Nature Musings from a Car Window," she wrote: "It goes without saying that the 'Nature-studyer' is sure to find more of interest in the change of scenes that meet the eyes of the traveler than is possible to one who has not learned to see what he looks at" (301). The lessons of nature study, then, could enhance the experiences of adults on vacations, as well as children on the farm. From a train window, the traveler could see "that self-satisfied little rodent, the prairie dog. . . . The little rascal has learned that trains are harmless and so he sits upright on his tripod of hind legs and tail, holding his forefeet drooping affectedly, while he gazes at us with as much interest as we in him" (301).

In her essays for adults in periodicals such as *Nature Magazine*, Comstock attributed human sentiments even to the plant world. In describing a dead Cholla cactus, she noted, "the branches, minus many of the terminal joints, hang down dejectedly; someone suggested that its general appearance of despondency was caused by contrition for having lived all its life a Cholla," inflicting its numerous and painful spines on all who came near it ("Cholla" 45–46). The moral lessons, anthropomorphizing, and creation of a relationship between animal and human were still present but aimed at a more mature audience in these essays.[4]

Comstock's popularity was built on a melding of accurate science with popular sentimentality and her aesthetic talents, as she sought to celebrate that sense of the beauty of nature through both words and images. In all her writings, Comstock incorporated nature poetry and essays by others. In her texts, describing the anatomy and life habits of plants and animals, she wove the verses she had come to love as a child, from the ancient Greeks to Byron, James Whitcomb Riley, Elizabeth Akers, William Cullen Bryant, Ralph Waldo Emerson, Henry David Thoreau, and Walt Whitman.[5] For example, she included a John Keats poem in an essay on the grasshopper:

> The poetry of earth is never dead;
> When all the birds are faint with the hot sun,
> And hide in cooling trees, a voice will run
> From hedge to hedge about the new-mown mead;
> That is the grasshopper's. He takes the lead
> In summer luxury. (*Ways of the Six-Footed* 14–15)

A similar pattern of artistic growth can be seen in Comstock's visual representations. After she mastered the stylized approach of scientific illustration—stark, unadorned representations of important anatomical features (figure 7.3)—Comstock began to develop an aesthetic interpretation of the natural world, producing scenes of the countryside from forests to brooks to moths against the moonlight (figure 7.4). In these works, she merged her artistic sensibilities with her knowledge of the natural world and her technical skills as an engraver to produce accurate yet subjective and aesthetic interpretations of the natural world. While technically accurate, in figure 7.4 the luna moth is dramatically cast against the moonlight and shadowy trees. Both the forms and the color contrasts create a memorable interpretation of an every-

Figure 7.3. *Marumba modesta.* From *Manual for the Study of Insects.*

Figure 7.4. Luna moth. From *Manual for the Study of Insects.*

day event. The overall composition is designed to capture the beauty and surreal aspects of nocturnal life.

In works like the *Manual*, Comstock's engravings broke out of the constraints of scientific illustration to include six plates that moved beyond technical drawing to aesthetic interpretations of the natural world (figure 7.5). The Comstocks' volume differed in this way also from other scientific textbooks of the day, such as Packard's, which were scantily illustrated and often relied on poor copies of illustrations produced a century or more earlier (Meadows 23–25). In both her writings and her illustrations, then, Comstock brought to their joint publications her own view of the natural world, a view that reflected her role as a woman and artist.

In her more popular works, Comstock gave free play to her artistic skills, making the pages of her essays an aesthetic delight. In many of her publications, each chapter opened with an illuminated letter that illustrated the theme of that chapter. In figure 7.6 the illuminated letter *L* echoes the written instructions to "Look at an insect. . . ." In other volumes, illustrations frame the text on the page, conveying both an accurate representation of the life forms being studied and Comstock's sense of the great beauty in nature. Figure 7.7 shows how a chapter entitled "Orchard Life" in John Henry Comstock's *Insect Life* (1897) opened with a beautiful but accurate depiction of knotgrass and the knotgrass beetle woven into the page design. Even after she gave up the time-consuming work of illustrating the Comstocks' publications, she continued to use her earlier engravings and hired a fine group of illustrators at Cornell, W. C. Baker, O. L. Forbes, and Louis Agassiz Fuertes, to produce original engravings. When halftone photography became readily available, she also used this medium for both its exact reproduction and its aesthetic possibilities. She thus ensured the continued quality of visuals in their publications.

In this role of interpreter, Comstock was not required to assume an "objective" stance toward the natural world required by western science, but rather was able to incorporate the aesthetic and emotional relationship to the natural world that was such an integral part of her personality (Keller 75–85). Comstock was first and foremost an artist, viewing nature in subjective terms. She expressed her views of nature in such symbolic media as poetry and the visual arts. Since childhood she had delighted in the symbolic language of such fellow artists as Keats, Wordsworth, and Byron. Their poetry conveyed an emotional relationship to nature and compressed complex relationships into symbolic forms. As an adult, she popularized these artistic works with a broad audience and mastered these artistic forms herself. Thus art, not

Figure 7.5. Plate VI. From *Manual for the Study of Insects*.

PART I.

LESSONS IN INSECT LIFE.

CHAPTER I.

THE PARTS OF AN INSECT.

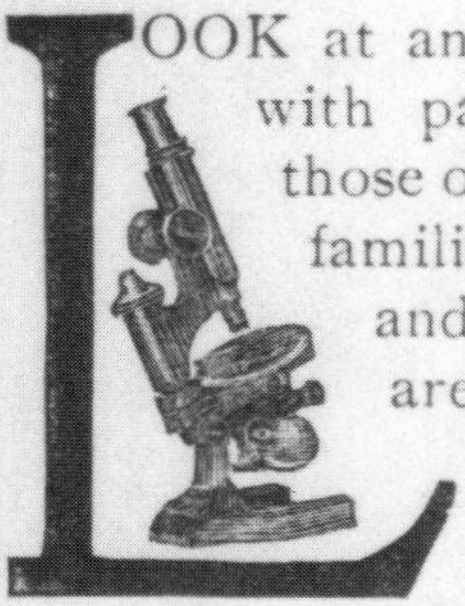

LOOK at an insect and you will find a creature with parts which are very different from those of the animals with which we are more familiar. Although it can see, hear, eat, and walk, its eyes, ears, mouth, and legs are not like the corresponding organs of the higher animals.

It is necessary, therefore, at the beginning of our study of insect life, to learn something of the structure of insects. We will not attempt at first, however, to make a thorough study of insect anatomy, but will merely select one kind of insect, and study the principal divisions of the body as seen from the outside.

Having done this, we will be able to see in our later studies in what ways the parts of other kinds of insects have been modified in form to fit them for their modes of life. Thus, for example, we will find that an insect which catches its prey by running has legs of a different shape than those of an insect that

9

Figure 7.6. "Look at an insect . . ." From *Insect Life*.

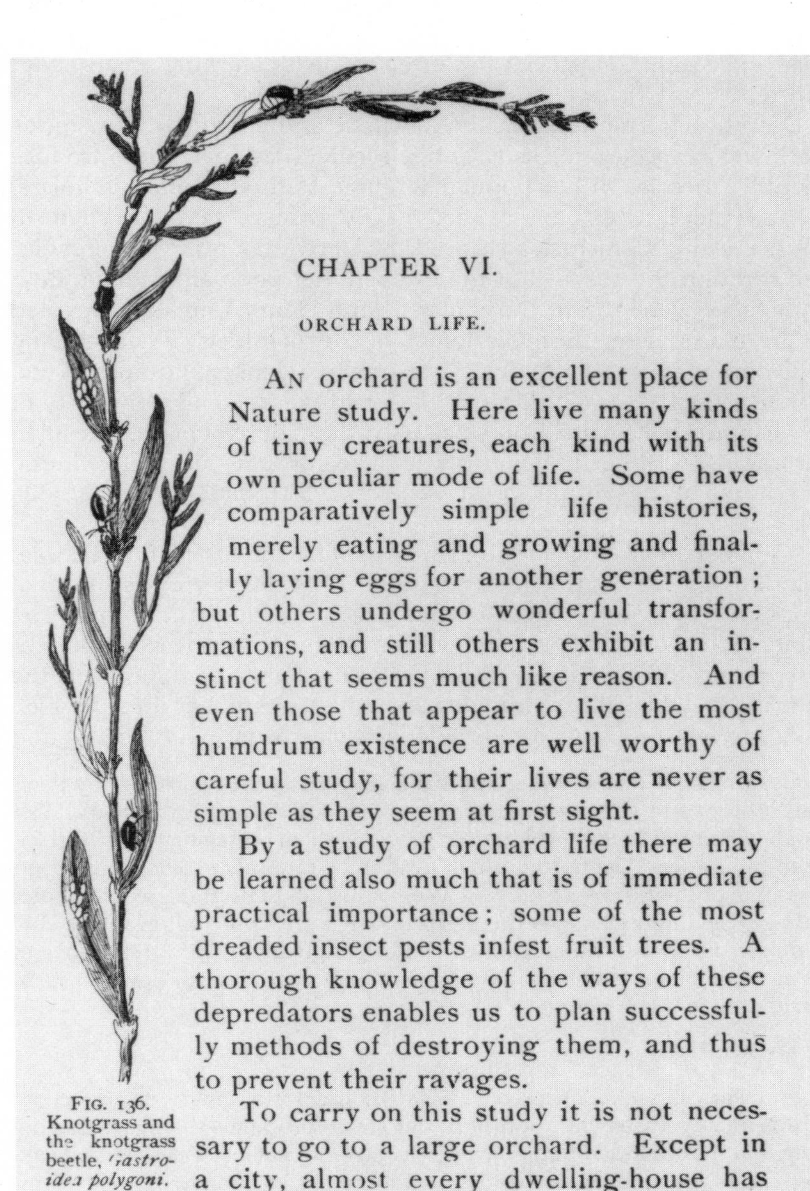

CHAPTER VI.

ORCHARD LIFE.

An orchard is an excellent place for Nature study. Here live many kinds of tiny creatures, each kind with its own peculiar mode of life. Some have comparatively simple life histories, merely eating and growing and finally laying eggs for another generation; but others undergo wonderful transformations, and still others exhibit an instinct that seems much like reason. And even those that appear to live the most humdrum existence are well worthy of careful study, for their lives are never as simple as they seem at first sight.

By a study of orchard life there may be learned also much that is of immediate practical importance; some of the most dreaded insect pests infest fruit trees. A thorough knowledge of the ways of these depredators enables us to plan successfully methods of destroying them, and thus to prevent their ravages.

To carry on this study it is not necessary to go to a large orchard. Except in a city, almost every dwelling-house has

Fig. 136. Knotgrass and the knotgrass beetle, *Gastroidea polygoni.*

166

Figure 7.7. "Orchard Life." From *Insect Life*.

science, became Comstock's preferred mode for teaching others about nature.

Despite her emphasis on the aesthetic and personal, Comstock's work was respected for its scientific accuracy, and she did introduce scientific theories in her popular writings. Although many scholars, such as Peter Bowler, have claimed that Darwinism was out of vogue in this era, Anna Comstock advanced the Darwinian principle of evolution through the survival of the fittest in her popular writings. Both Comstocks were ardent Darwinians. John Henry Comstock devoted his life to unraveling the evolutionary history of insects and developing an evolutionary methodology for taxonomy (Henson, "Comstock Research School" 161–67). Despite the controversies and popularity of neo-Lamarckism, Anna Comstock used the "survival of the fittest" in her popular writings to demonstrate the complexity, interrelatedness, and drama of the natural world (*Ways of the Six-Footed* vii, 39–53, 130–31).

"A Sheep in Wolf's Clothing" in *Ways of the Six-Footed*, for example, describes mimicry of the monarch butterfly by the viceroy as a form of adaptation by natural selection. She tells her young reader that "The habits of a species comprise the wisdom stored up in the experience of that species during thousands of years; the habits of a species is the pathway by which it has struggled up to the ranks of the "fittest" which have survived" (39). As a caterpillar, the viceroy is

> a most grotesque and amazing-appearing creature, with a pair of spiny pompons in front and spines too numerous to mention decorating his body. Most people not entomologically educated would exclaim on seeing this caterpillar when full grown "The horrid thing!" And if the caterpillar could hear and be conscious of the history of his race as embodied in himself, he would rejoice and be exceedingly glad over this verdict; for it is greatly to his advantage now to look so disagreeable that no one would willingly molest him. The height of his racial ambition is to be so humpy and spiny that no bird, however rash, would dare to touch him. (44)

As an adult,

> Its gay color, orange-red marked with black borders and veins, is its protection; for it is an advertisement a sort of poster which proclaims that here is something that right-minded birds leave alone. So our palatable Viceroy has developed colors and markings so nearly like the unpalatable Monarch that no feathered creature will touch him. (48–49)

Throughout this essay, Comstock refers to the history of the species, adaptation through camouflage to surroundings and other species, and the survival of the fittest. Like those of many Darwinians, her sto-

Figure 7.8. John Henry Comstock, courtesy of the Charles P. Alexander Papers, Smithsonian Institution Archives.

ries are teleological, that is, they suggest that the animal purposefully changed in an adaptive way, rather than owing to random forces. But she is clearly inculcating a Darwinian point of view and teaching children to find fascination even in caterpillars that lack beauty.

Although Comstock taught science in her essays, she did not believe that children needed to master the complex language of objective science, including Linnaean nomenclature for plants and animals. Like earlier women nature writers, she advocated using common names when teaching young people about the natural world (Wakefield 30–31). Although she often gave the Latin binomials, she always used the common name. She insisted that children not be turned against science by requiring rote memorization of long and strange terms. Science was part of living, not a temple guarded by high priests with a specialized liturgy. She noted, however, that "most children like a 'word that is a mouthful' " ("Should We Use Scientific Names?" 51); thus, they should not be discouraged from learning complex terms when interested. Many a nature study student, she knew, had gone on to a career in natural history, and such study built a firm foundation for budding naturalists (V. Bailey).

Despite her expertise, Comstock never presented herself to the reader as an accomplished scientist. She was frequently at pains to point out her limitations as a science interpreter rather than researcher. She modestly noted that she was drawing on the work of well-known scientists, complementing their scholarly tomes with her personal observations of plants and animals (*Ways of the Six-Footed* vii–viii). As Kohlstedt and Shteir have noted, this modesty was typical of women nature writers, perhaps to deflect criticism that they had become too intellectual or knew too much.

Although Comstock strove for scientific accuracy, her language was often sentimental when describing sexual organs and behavior, invoking romantic, marital, and maternal analogies. Describing the mating of watersprites, she wrote,

> His life in the air is short and sweet. Hiding himself during the garish day he comes out in the shadowy night and seeks his mate. She may have been his nearest neighbor at the bottom of the stream, but she was nothing to him then. Now she is all there is of life's happiness. As soon as he, after brief possession, loses her, he seems to realize there is naught else worth living for, and he dashes toward the first light that affords him the opportunity for self-immolation. (*Ways of the Six-Footed* 137–38)

Since Linnaeus's day, explicit references to sexual organs and behavior had been viewed by many as an obstacle to women pursuing science.

The result was bowdlerized texts that would protect women from such language (Shteir, "Linnaeus's Daughters" 73). Even Darwin, in his *The Descent of Man, and Selection in Relation to Sex,* used romantic terminology to describe mating. When discussing the evolution of the cicada's mating calls, he wrote, "A grating sound thus occasionally and accidentally made by the males, if it served them ever so little as a love-call to the females, might readily have been intensified through Sexual Selection, by variations in the roughness of the nervures having been continually preserved" (297). Thus Comstock was working within an established tradition of using romantic language to describe mating in scientific and popular writings. And indeed, when she did refer to mating, she couched it in sentimental terms calculated to make it less threatening to naive or prudish readers. It was precisely this type of language that later brought the nature study movement under attack as unscientific.

Although Comstock did not teach laboratory biology, nature study entailed collecting and dissecting organisms. Comstock tried to minimize the amount of destruction inherent in such study. She emphasized observation in the field, and in her more popular essays would assure her readers that she had returned the somewhat ruffled creature to his or her home in the woods or pond. When collection and dissection were required, her husband advised,

> In doing this we should be humane. It is not probable that insects are as sensitive to pain as we are, but there is no doubt that they suffer when injured. We should, therefore, handle our specimens carefully, kill them without inflicting needless pain, and destroy no more than is absolutely necessary for study. It is not merely the insects that are to be considered in this matter, for no one can be cruel to animals without its having a bad effect on his character. (J. H. Comstock, *Insect Life* 284)

Such concerns about the salutary versus depraving effects of animal collecting on the character of the collector had generated heated debates in England earlier in the century. One poet declared,

> For who'er to a fly can barbarity show
> Will not scruple the worst deed to do.[6]

Oddly enough, much of this attention focused on the destruction of insects, since the killing of birds and mammals was sanctioned by the English hunting tradition (Larsen 90–127). Given the moral lesson nature study sought to teach, its proponents certainly did not want to be accused of teaching cruelty to young people; thus, detailed instructions were provided to ensure that the killing of organisms was done as quickly and humanely as possible (J. H. Comstock, *Insect Life* 24–25, 284–93).[7]

Comstock's sense of connectedness to the natural world can be seen in the many lessons she taught about responsibility for the natural environment. She often pointed out the many benefits offered by animals or plants, from recycling to pest control to maintaining the balance of nature (*Handbook of Nature Study* vii). In her *Handbook* she opened the essay on the bumblebee with a discussion of the "hereditary war between the farm boy and the bumblebee" (389–91), noting that it had caused great harm to both sides. In this way, she created an immediate personal relationship between the reader and the bee, as well as a sense of responsibility for the bumblebee's fate. She pleaded for their preservation, based on their contributions to agriculture as pollinators. But she also argued that nature must be loved and enjoyed for its intrinsic value. Comstock still had not lost her respect for and joy in the beauty of the natural world and fascination with the marvelous life habits and adaptations to be found in every woods and pond. Indeed, her lessons of respect for the beauty and harmony of nature, for its complexity and interrelatedness, are often credited with having a significant influence on the generation of environmentalists who followed her.

Although Anna Botsford Comstock entered science through the "back door" by assisting her husband, she then used her many talents and considered views on women, science, and education to create an influential and satisfying career for herself. An excellent observer and synthesizer, she also brought a personal and aesthetic approach to nature that challenged the accepted "objective" and analytical stance. Thus her work was a unique blend of accurate science, aesthetic appreciation of nature, visual literacy, and learning theory that emphasized direct experience of nature. Comstock drew on the work of her predecessors and was deeply influenced by them, but she created her own synthesis, one that was compatible with her views about nature, gender relations, children, and aesthetics. Nature study proved the ideal vehicle for Anna Comstock because it allowed her to express her more artistic/subjective side, continue to learn about the natural world, and teach children and the general public a different point of view toward nature, not merely scientific facts.

Although trained in scientific research, Comstock found a place for herself at the periphery, in scientific illustration, where women were traditionally more welcome. Here she was able to develop her aesthetic talents while complementing her husband's career. At the same time, she established herself as a professional, which led to new challenges in children's education and popular writing. As she explored these other accessible niches—agriculture extension, children's writing, edu-

Figure 7.9. John Henry and Anna Botsford Comstock. Courtesy of the Charles P. Alexander Papers, Smithsonian Institution Archives.

cation, and popular science writing—she found they valued feminine perspectives. She then synthesized all of these aspects of the discipline of science into a large and public role that gave her international visibility and influence. Although Anna Comstock accepted the notion of separate spheres for each gender and chose gender-appropriate roles, she also expanded the limits of women's sphere in science, and in this way for a time helped women move in from the periphery, closer to the mainstream.

Anna Botsford Comstock (1854–1930): Major Works

Manual for the Study of Insects (1895) [coauthored with John Henry Comstock]
How to Know the Butterflies: A Manual of the Butterflies of the Eastern United States (1904) [coauthored with John Henry Comstock]
Ways of the Six-Footed (1903)
How to Keep Bees (1905)
Confessions to a Heathen Idol (1906)
Handbook of Nature Study (1911)
The Pet Book (1914)
Trees at Leisure (1916)
The Comstocks of Cornell: John Henry and Anna Botsford Comstock (1953)

Notes

1. For further discussion of this expanding role for women, see Kohlstedt, "Parlors" 434–39; Shteir, "Botany in the Breakfast Room" 32–33, "Botanical Dialogues" 311–13, and "Priscilla Wakefield's Natural History Books" 29–30, 32.
2. For additional discussions of opportunities for women in science writing, see Kohlstedt, "In from the Periphery" 86–89, 96, and "Parlors" 425–45; Rossiter, "Women's Work in Science" 391–95; and Shteir, "Botany in the Breakfast Room" 34–36, and "Linnaeus's Daughters" 70, 72–73.
3. The Comstocks' popular books included *How to Know the Butterflies* (1904), *How to Keep Bees* (1905), *Insect Life* (1897), *The Pet Book* (1914), *Trees at Leisure* (1916), and *Ways of the Six-Footed* (1903).
4. See, for example, in *Nature Magazine*, "Nature Study Review," "Green Tree in the Desert," "Nature Musings from a Car Window," "Giant Cactus or Suharo," "Bird Study in Honolulu," "The Candle-Nut Tree," and compare her two versions of the same story about wasps, "The Ingenuity of Ants and Wasps" in *The Chautauquan*, and "Two Mother Masons," in *Ways of the Six-Footed* 96–106, which was aimed at children.
5. See for example *Handbook* 53, 57, 64, 65, 123, 155, 241, 389, 411, 549, 638, 667, and *Ways of the Six-Footed* 6–10, 14–19, 21.
6. "Young Philomon Accused by His Sister of Cruelty," in *Poems on Various Subjects for the Amusement of Youth* (1789), quoted in Larsen 111.
7. See also poem by Christina Rossetti in Comstock, *Handbook* 325; and William Cowper, "Care for the Lowest," in Firth.

Works Cited

Bailey, Liberty-Hyde. *The Nature-Study Idea: An Interpretation of the New School-Movement to Put the Young in Relation and Sympathy with Nature*. 1903.

Bailey, Vernon Orlando. "How to Become a Naturalist." *Nature Magazine* 3 (1924): 361–62.
Bonta, Marcia Myers. *Women in the Field: America's Pioneering Women Naturalists.* 1991.
Bowler, Peter J. *The Eclipse of Darwinism: Anti-Darwinian Evolution Theories in the Decades around 1900.* 1983.
Comstock, Anna Botsford. "Bird Study in Honolulu." *Nature Magazine* 6 (1925): 52–53.
Comstock, Anna Botsford. "The Candle-Nut Tree." *Nature Magazine* 6 (1925): 118.
Comstock, Anna Botsford. "The Chicadee—The Snow-Storm." *Chautauquan* 38 (1903): 387–90.
Comstock, Anna Botsford. "Cholla, the Pest of the Desert." *Nature Magazine* 4 (1924): 456.
Comstock, Anna Botsford. "The Clovers—The Bumblebee." *Chautauquan* 39 (1904): 384–90.
Comstock, Anna Botsford. *The Comstocks of Cornell: John Henry and Anna Botsford Comstock.* 1953.
Comstock, Anna Botsford. *Confessions to a Heathen Idol.* 1906.
Comstock, Anna Botsford. "The Evergreens." *Nature Study Review* 2, no. 1 (1906): 1–13.
Comstock, Anna Botsford. "Geography and Life." *Nature Study Review* 16 (1920): 75–80.
Comstock, Anna Botsford. "Giant Cactus or Suharo." *Nature Magazine* 3 (1924): 364–65.
Comstock, Anna Botsford. "Green Tree in the Desert." *Nature Magazine* 3 (1924): 234–35.
Comstock, Anna Botsford. *Handbook of Nature Study.* 1911.
Comstock, Anna Botsford. "Hermit of Dyer's Pond." *Nature Study Review* 19 (1923): 164–67.
Comstock, Anna Botsford. "Hidden Homes." *St. Nicholas* 16 (1889): 605–7.
Comstock, Anna Botsford. "How to Begin Bee-Keeping." *Country Life in America* 7 (1905): 636–38.
Comstock, Anna Botsford. *How to Keep Bees.* 1905.
Comstock, Anna Botsford. "Industries of the Hive: Secreting Wax and Comb Making." *Country Life in America* 8 (1905): 462–64.
Comstock, Anna Botsford. "The Ingenuity of Ants and Wasps." *Chautauquan* 26 (1898): 590–93.
Comstock, Anna Botsford. "Insect Communities." *Chautauquan* 26 (1898): 479–83.
Comstock, Anna Botsford. "Insect Domestic Economy." *Chautauquan* 27 (1898): 294–98.
Comstock, Anna Botsford. "Insect Musicians." *Chautauquan* 27 (1898): 652–59.
Comstock, Anna Botsford. "Nature Musings from a Car Window." *Nature Magazine* 3 (1924): 301–2.

Comstock, Anna Botsford. "Nature Study: Beginning Bird Study." *Chautauquan* 41 (1905): 259–64.

Comstock, Anna Botsford. "Nature Study: The Evergreens." *Chautauquan* 40 (1904–5): 366–69, 465–68.

Comstock, Anna Botsford. "Nature Study: Seed Distribution." *Chautauquan* 40 (1904): 271–73.

Comstock, Anna Botsford. "Nature Study: Tree Study in Winter." *Chautauquan* 41 (1905): 66–72.

Comstock, Anna Botsford. "Nature-Study and Agriculture." *Nature Study Review* 1, no. 4 (1905): 143–47.

Comstock, Anna Botsford. "Nature Study Review." *Nature Magazine* 3 (1924): 45.

Comstock, Anna Botsford. "Observation Bee-Hive for the Schoolroom." *Nature Study Review* 1, no. 3 (1905): 109–14.

Comstock, Anna Botsford. "Our Own Sun and His Own Family." *Nature Study Review* 18 (1922): 305–16.

Comstock, Anna Botsford. *The Pet Book.* 1914.

Comstock, Anna Botsford. "The Ripened Corn—The Ways of the Ant." *Chautauquan* 38 (1903): 172–77.

Comstock, Anna Botsford. "Should We Use Scientific Names?" *Nature Magazine* 5 (1925): 51.

Comstock, Anna Botsford. "Social Habits of Insects." *Chautauquan* 26 (1898): 366–69.

Comstock, Anna Botsford. "The Sugar Maple—The Red Squirrel." *Chautauquan* 38 (1903): 281–86.

Comstock, Anna Botsford. "Suggestions for a Graded Course in Bird Study." *Nature Study Review* 16 (1920): 147–58.

Comstock, Anna Botsford. "Trees at Leisure." *Country Life in America* 1 (1902): 105–9.

Comstock, Anna Botsford. *Trees at Leisure.* 1916.

Comstock, Anna Botsford. *Ways of the Six-Footed.* 1903.

Comstock, Anna Botsford. "The White-Breasted Nuthatch—Our Use of Food Stored in Seeds." *Chautauquan* 38 (1904): 491–96.

Comstock, Anna Botsford. "Whitefoot, as a Boarder." *Nature Study Review* 19 (1923): 240–42.

Comstock, John Henry. *Insect Life: An Introduction to Nature Study.* 1897.

Comstock, John Henry. *Introduction to Entomology.* 1888.

Comstock, John Henry, and Anna Botsford Comstock. *How to Know the Butterflies: A Manual of the Butterflies of the Eastern United States.* 1904.

Comstock, John Henry, and Anna Botsford Comstock. *Manual for the Study of Insects.* 1895.

Conable, Charlotte Williams. *Women at Cornell: The Myth of Equal Education.* 1977.

Darwin, Charles. *The Descent of Man, and Selection in Relation to Sex.* 1989.

Fenn, Lady Elleanor. *Short History of Insects.* 1797.

Firth, Abraham, ed. *Voices for the Speechless.* Boston: Houghton Mifflin, 1883.

Ford, Laetitia. *The Butterfly-Collector's Vade Mecum*. 1824.
Henson, Pamela M. "The Comstock Research School in Evolutionary Entomology." *Osiris* 8 (1993): 159–77.
Henson, Pamela M. "The Comstocks of Cornell: A Marriage of Interests." In *Creative Couples in the Sciences*, ed. Helena M. Pycior et al. New Brunswick: Rutgers UP, 1996. 112–25.
Henson, Pamela M. "Evolution and Taxonomy: John Henry Comstock's Research School in Evolutionary Entomology at Cornell University, 1874–1930." Diss., University of Maryland, 1980.
Keller, Evelyn Fox. *Reflections on Gender and Science*. 1980.
Kohlstedt, Sally Gregory. "In from the Periphery: American Women in Science, 1830–1880." *Signs* 4 (1978): 81–96.
Kohlstedt, Sally Gregory. "Parlors, Primers, and Public Schooling: Education for Science in Nineteenth-Century America." *Isis* 71 (1980): 381–98.
Larsen, Anne L. "Not since Noah: The English Scientific Zoologists and the Craft of Collecting, 1800–1840." Diss., Princeton University, 1993.
Mallis, Arnold. *American Entomologists*. 1971.
Meadows, A. J. "The Evolution of Graphics in Scientific Articles." *Publishing Research Quarterly* 7 (1991): 23–32.
Packard, Alpheus Spring. *Guide to the Study of Insects, and a Treatise on Those Injurious and Beneficial to Crops*. 1869.
Palmer, E. Laurence. "The Cornell Nature Study Philosophy." *Cornell Rural School Leaflet 38* (1944): 3–80.
Rossiter, Margaret W. *Women Scientists in America: Struggles and Strategies to 1940*. 1982.
Rossiter, Margaret W. "Women's Work in Science, 1880–1910." *Isis* 71 (1980): 381–98.
Shteir, Ann B. "Botanical Dialogues: Maria Jacson and Women's Popular Science Writing in England." *Eighteenth-Century Studies* 23 (1990): 301–17.
Shteir, Ann B. "Botany in the Breakfast Room: Women and Early Nineteenth Century British Plant Study." In *Uneasy Careers and Intimate Lives: Women in Science, 1789–1979*, ed. Pnina Abir-Am and Dorinda Outram. 1987.
Shteir, Ann B. "Linnaeus's Daughters: Women and British Botany." In *Women and the Structure of Society*, ed. Barbara J. Harris and JoAnn K. McNamara. 1984.
Shteir, Ann B. "Priscilla Wakefield's Natural History Books." In *From Linnaeus to Darwin: Commentaries on the History of Biology and Geology*, ed. Alwyne Wheeler and James B. Price. 1985.
Treat, Mary. *Home Studies in Nature*. 1885.
Wakefield, Priscilla. *Introduction to the Natural History and Classification of Insects*. 1816.
Whalen, Matthew D., and Mary F. Tobin. "Periodicals and the Popularization of Science in America, 1860–1910." *Journal of American Culture* 3 (1980): 195–203.

PART 5

DEFINING AND REDEFINING KNOWLEDGE: POST-DARWINIAN WOMEN

8

Revising the Descent of Woman
Eliza Burt Gamble

Rosemary Jann

The rise of evolutionary thinking in the nineteenth century enabled a powerful new genre of origin myths for human life, myths that replaced the old authority of religion with the new sanctions of scientific law. Victorian ethnologies aimed to subsume the exotic past into a master narrative of western progress, while biological accounts aimed to give human nature a history. For both, scientific explanation, like earlier forms of mythmaking as well as current practice,[1] was a kind of storytelling, a hypothetical reconstruction of the processes that led to present realities. Both were implicitly allegorical, in the sense that they assumed and helped to construct a transcendent level of truth about human life that gave meaning to their narrative ordering.[2] Crucial to both kinds of history were explanations (and implicitly, justifications) of gender difference, something we still view as one of the deepest truths of human identity. Nineteenth-century origin accounts imagined certain family structures and sexual relations as crucial to defining the border between animal and human nature and explaining the development of human civilization.

Given the teleological bias of most historical narratives, it was almost inevitable that in the process such accounts would write back into history the realities of social power in the present, assuming these as human universals in need of explanation, so that the Victorian sub-

Figure 8.1. Eliza Burt Gamble (1841–1920). Reproduced courtesy of the *National Cyclopedia of American Biography* from the Collections of the Library of Congress.

ordination of women became a "fact" of nature and a pillar of civilization. Evolutionary biology shared with social anthropology a concern to replace supernatural intervention with naturalistic laws whose uniform operation could account for present human order. Darwin accomplished this in *The Descent of Man and Selection in Relation to Sex* by hypothesizing ways in which biology could have produced culture. Social evolutionists turned to the comparative method to construct a unilinear scale of progress upon which all peoples, living and dead, could take their place. Crucial to both explanatory models were similar assumptions that made male agency and male control of women the engine of human progress. The second half of the century witnessed a veritable explosion of similar efforts to justify female inferiority on "scientific" grounds.[3]

Among contemporaneous feminist responses to such theories, the work of American suffragette Eliza Burt Gamble is distinctive for its insights into the ways the biological and anthropological rationalizations of female inferiority served one another and for its detection of the logical inconsistencies that both depended upon. Gamble's *The Evolution of Woman* (1894), revised and expanded as *The Sexes in Science and History* (1916),[4] focused on strategic discontinuities between animals and humans in these accounts and responded with a counternarrative that rewrote female power over sexuality as the missing link between biology and culture. Although no more able than her opponents to escape from essentialist conceptions of sexual identity, Gamble was able to achieve shrewder insights into the ways culture and economic power constructed female weakness, at least in part because her socialist sympathies did not commit her to justifying modern property relations as the highest development of human civilization, as was the case with many Victorian ethnologists.

Eliza Burt was born in Concord, Michigan, in 1841. Left an orphan at sixteen, she taught in various Concord public schools and was a superintendent in East Saginaw. She became active in the women's suffrage movement shortly after her marriage to lawyer James Gamble in 1867, helping to organize Michigan's first women's suffrage convention in 1876. A frequent contributor to newspapers in New York and the Midwest, she turned away from suffrage activities in the eighties to research the historical roots of women's oppression, publishing *The God-Idea of the Ancients,* which considered the ways patriarchal thought had shaped religions, in 1897. An invalid for the last twenty years of her life, she died in 1920 and was survived by two of her three children (*NCAB* 18:220–21; *Detroit Free Press,* 19 Sept. 1920, 1:6).

Gamble's approach to that body of theory that Cynthia Russett has labeled "sexual science" reminds us how wide was the umbrella of evolutionary thought in the late nineteenth century, when fine distinctions were seldom made between Darwin's ideas (and their legitimate applications) and the wide array of other developmental hypotheses offering biological explanations for cultural phenomena. Although her investigation of woman's place in nature was originally inspired by reading *The Descent of Man* (*SSH* v), *The Sexes in Science and History* adopted a rather Lamarckian conception of adaptation, in which inherited forms became modified "so as to perform more complicated functions, or those better suited to their environment" (*SSH* 8), and a Spencerian equation of greater differentiation with higher specialization (*SSH* 14). Rather than accepting the common scientific assumption that woman was less differentiated than man, having evolved less beyond the young of the species (Russett 74–75, 92–93), Gamble argued that the female reproductive organs, developed to "nourish and protect the germ," were more highly specialized than the male's and thus represented a higher stage of development (*SSH* 15). She went on to moralize evidence of sexual differentiation by finding in the contrast between male traits to catch and hold the female, and female traits to nurture the young, the biological origin of that altruism that made the female morally superior to the selfish male (*SSH* 16–17), thus providing evolutionary backing for a distinction commonly made by nineteenth-century feminists (Hersh 274). Her analysis in this sense differed significantly from that of Antoinette Brown Blackwell's earlier feminist reply to Darwin, *The Sexes throughout Nature* (1875), which stressed the complementarity of the sexes and their separate but equal contribution to nurturing the young.[5]

Darwin's hypothesis of sexual selection was central to the *Descent*, and Gamble makes it central to her analysis of sexual difference as well. Darwin had devised this theory to explain differences between male and female that could not be attributed to natural selection, arguing that those males with stronger weapons to drive off rivals and/or more distinctive markings to attract females would have greater reproductive success, thus increasing the incidence of these traits among their male offspring. Unlike many of Darwin's contemporaries (Russett 89–92), Gamble endorsed his explanation of sexual differentiation, but interpreted it in such a way as to challenge many of the androcentric assumptions based upon it. Darwin's theory predicted, and evidence from animals confirmed, that males were more variable in structure than females (Darwin 580–81), and this variability had encouraged the development of more distinctive secondary sexual characteristics. Their greater differentiation prompted Darwin to treat men as being further evolved than women, a position that at least tacitly supported later claims that females were in effect "arrested males," as Johns Hopkins zoologist W. K. Brooks taught (Russett 75, 93). Rejecting this linear scaling of human development, Gamble instead focused on the modeling of specialization and force, arguing that those highly variable male secondary sexual characters denoted insufficient specialization and hence lower organization than the female's (*SSH* 30). She stressed instead the dysfunctional, even "injurious," nature of these characters—wings too large to fly, antlers so heavy as to retard escape from predators—and the greater waste of "vital force" needed to create them, force that, according to her Lamarckian conception of adaptation, might otherwise have been used to perfect the organism or better fit it to its environment (*SSH* 30–31, 35; cp. Darwin 581). Gamble took Darwin's evidence of the greater incidence of insanity, reversion, and disease in males as further proof of their physiological inferiority to females (*SSH* 45–51).[6]

Patrick Geddes and J. Arthur Thomson's more static and essentialist explanation in *The Evolution of Sex* was in some ways better suited to Gamble's argumentative needs than Darwin's dynamic model; she cited it to claim that conditions of greater abundance favored the production of females, just as higher nutrition was needed to convert a drone into a queen bee (*SSH* 39–40). But where Geddes and Thomson's metabolic model had associated "katabolic" male activity with greater intelligence and originality, and "anabolic" female passivity with a placid and quiescent conservatism (Geddes and Thomson 270–71), Gamble argued instead that precisely because the female did not waste force producing maladaptive sexual traits, what she actually conserved was "all the gain which has been effected through Natural Selection" (*SSH* 30), and not just the lowest common denominator of bodily form. In reply to longstanding misogynist arguments that treated females as mere vessels to facilitate male creative powers, Gamble opined instead that "it would seem that the female is the primary unit of creation, and that the male functions are simply supplemental or complementary" (*SSH* 38).[7]

Gamble similarly contested the conclusions Darwin drew about female choice in sexual selection. He had theorized that among animals, the greater distinctiveness of the male form resulted from the female's choosing those individuals whose beauty pleased her most. Whereas the male was eager for any female, the female was selective and coy, needing to be courted before accepting a mate—much like Victorian maidens.[8] Although near the end of the *Descent* Darwin linked the exertion of sexual choice to the development of higher mental powers (Darwin 918), when considering women he tended to dismiss their characteristic mental traits—perception and intuition—as typical of lower races and earlier stages of civilization (Darwin 873). Gamble opted for a more consistent extension of his logic, basing claims for superior female intelligence on the powers of discrimination, taste, and beauty that must have been developed by their choice of animal mates (*SSH* 29). She shrewdly identified the issue in sexual selection that has attracted feminist criticism ever since: the shift from female choice in animals to male choice in people. Woman's greater beauty convinced Darwin that men must have gained the power of selection; although he might allow that in civilized societies, women have "free or almost free choice" of husbands (Darwin 891), in practice he talks almost exclusively of the effects of male choice in humans, the most important of which historically speaking was female subordination to male power. Although Darwin might deplore male selfishness and pay lip service to the importance of maternal altruism in extending human sociability beyond the family (Darwin 873), in practice he treated male competition as vital to human progress and female dependency as crucial to the forging of human society. Whenever Darwin imagines our semihuman ancestors, he envisions a patriarchal family group in which the males are already choosing, controlling, and jealously guarding their mates (e.g., Darwin 430, 898–99, 907–8). Through an elision of the difference between sexual and natural selection, he assumed that male control necessarily entailed female dependency and that this dependency in turn explained the mental and physical inferiority so apparent in modern woman, for once the female no longer had to struggle to defend and feed herself, she would no longer have been subject to natural selection for strength and intelligence (Russett 80–84; Richards 71–73). It was fortunate, he added, that among mammals male intellect could be transmitted equally to both sexes; "otherwise it is probable that man would have become as superior in mental endowment to woman, as the peacock is in ornamental plumage to the peahen" (Darwin 874).

Adopting a model of human evolution that took altruism rather than control as its motive force, Gamble again used the logic of Darwin's argument to challenge the conclusions drawn from it. Stressing the importance he accorded the social instincts in developing morality, she argued that these could only have developed from maternal sympathy and altruism. The male pugnacity and perseverance favored by sexual selection did not encourage a concern for the welfare of others; animal evidence made clear that even among one's own offspring, paternal affection was virtually unknown until very recently in evolutionary history (*SSH* 69–71). Darwin's model assumed that as soon as human beings appeared, the "processes which for untold ages" had operated to develop superior powers of female sympathy and discrimination were reserved, allowing males to gain control of progress through the pursuit of essentially animal gratification. It would be more consistent to argue that without Darwin's law of equal transmission of characters to both sexes, man would never have been able to rise above his animal drives and inherit the social instincts that eventually made moral behavior and advanced human society possible (*SSH* 74–76).

Some of the inconsistencies in Darwin's argumentation were linked to his difficulty in grafting a biological explanation of the origins of culture onto the analyses of early human life that he adopted from Victorian ethnologists.[9] It was crucial to the *Descent*'s argumentative logic that Darwin be able convincingly to represent the behavior of the earliest humans as motivated by the same principles that governed animal sociability. And yet, after having cast animal mating behavior in the anthropomorphized terms of Victorian courtship and marriage, Darwin found himself having to explain the patently "unnatural" sexual behavior of primitive man that he adopted from the reconstructions of Victorian ethnologists like John Ferguson McLennan and John Lubbock. As George Stocking has explained, such social evolutionists aimed to refute claims mounted by religious conservatives that humankind could not have made the step from savagery to civilization without supernatural help, and thus that modern savages must have degenerated from higher levels. Assuming that social development, like the rest of the natural world, could be explained by uniform laws, they postulated a single scale of development for human civilization, treating modern savages as representative of what all early societies must have been like (Stocking 169–70). They extrapolated backward from modern man, reconstructing a series of steps leading from the anarchic licentiousness of savage sexual relations to the patriarchal family that, in their view, had made possible the sublimation of animal desire in the companionate marriages and parental affection of the modern day.

Their backward projections, however, stopped short of the border between animal and human, leaving a gap that Darwin had to negotiate in his own reasoning forward from animal to man (Stocking 176–78). Elizabeth Fee has argued that the anthropologists treated savage promiscuity as the logical result of insufficient moral development; patriarchal control of sexuality was not natural, but rather the hard-earned fruit of social progress (Fee 89–90, 94). Having already imagined hominid families as controlled by jealous fathers, and having taken filial and parental affections in animals as the origin of human sociability and morality, Darwin faced in his dealings with savage behavior rhetorical obstacles to continuity that the anthropologists did not encounter. On the one hand, he needed to position modern savages as missing links between animals and civilized man to make convincing the gradual development of the higher faculties from animal instinct. But on the other, the evidence of infanticide, promiscuity, matriarchy, and polyandry in savage life represented an offensive reversal of what he had constructed "natural" sexuality to look like in hominids. Having left one rhetorical gap between female selection of males in lower animals and male choice of females in humans, he created another by arguing that increasing reason, elsewhere the mechanism for developing morality out of the social affections (Darwin 471–72), led in savages to a perversion of those same affections and a reversal of instinctive sexual behavior. Between the jealous patriarchy of "our early semi-human progenitors" (Darwin 430) and the enlightened marriages of modern Europeans, his natural history of human desire was thus ruptured by savage resistance to normative sex roles, and the advance into full humanity became a fall from natural instincts.

Gamble fully appreciated the rhetorical significance of these discontinuities in Darwin's narrative and offered her maternalist account as a more biologically consistent explanation of the transition from animal to human sexuality, as well as one that could expose the androcentric prejudices that led the anthropologists to make civilized progress dependent upon female subordination. Just as the *Descent* contains the subplot of a more natural age of hominid patriarchy at least temporarily overthrown by savage perversity, so too does Gamble's account depend upon a lost Eden. But hers was a golden age in which the female power to control sexual relations in nature extended into a relatively recent period of human history; the overthrow of that power represented not "a regular, orderly, and necessary step in the direct line of progress" (*SSH* 99) as the anthropologists argued, but a retrogression in both natural and civil order. Victorian ethnologists had constructed from the degraded sexual life of modern savages a historical prototype for universal primitive promiscuity; Gamble found their accounts marred by the prejudices of men who saw "lawlessness" and immorality in any customs that granted women independence of male control. She took a far more utopian view of primitive life, arguing that genuine evidence of lewd behavior in modern savages was more likely to represent the corruptions of civilization and pointing to the strict regulation of sexual relations among many tribes (*SSH* 110–15). Gamble drew support from Darwin's belief that widespread primitive promiscuity was unlikely given animal precedent (*SSH* 115–16) in order to contest the anthropologists' primitive female, incapable of sexual restraint. But Darwin's position was based on what he saw as the universal jealousy of all male animals (Darwin 895). Gamble, on the other hand, universalized female virtue, making it not the cultural product of male control, as the anthropologists claimed, but essential to her nature. "The fundamental bias of the female constitution" would have ruled out promiscuous sexuality, she argued, since both animal and human evidence showed that "any degree of affection for more than one individual at the same time is contrary to the female nature" (*SSH* 107–8). In Gamble's evolutionary model, Victorian assumptions about woman's weaker and more family-oriented sexual instincts empowered a gynecocratic history of human civilization.

The anthropologists, on the other hand, plotted social evolution as a series of adjustments to natural depravity, progressing away from an aboriginal promiscuity and toward the increasingly successful constraints of the civilized world. The most radical explanation for matrilineal societies, for instance, came from the Swiss scholar Johann Jakob Bachofen, who argued that female disgust at primeval promiscuity had led to an Amazon overthrow that enforced matriarchy upon men. More sober English commentators like John McLennan and John Lubbock dismissed the idea that there had ever been a time when women exercised supreme power, for, as Lubbock pointed out, "we do not find in history . . . that women do assert their rights, and savage women would, I think, be peculiarly unlikely to uphold their dignity in the manner supposed" by Bachofen (104). They argued instead that given the obvious moral laxity of savages, descent in early times would have had to be reckoned matrilineally, since paternity would have been so uncertain during early periods of communal marriage (McLennan 64–70). C. Staniland Wake and Herbert Spencer similarly treated matrilineality as a mere convenience, conferring no real governing power on women (*SSH* 134).

For Gamble matrilineality and matriarchy represented the logical—and natural—extension of female control of sexual relations in the animal kingdom. She in effect took Darwin's equation of choice with control and applied it to females. Given that among lower animals the male must please the female to secure her favors, it was evident to Gamble that "under earlier and more natural conditions of human life, the appetites developed within him were still largely controlled by her will" (*SSH* 213). Gamble sketched a utopian view of early gens or kin groupings, in which all members traced their descent back to a single woman, egotism was unknown, and sympathy, "a sprout from the well-established root, maternal affection" (*SSH* 126), governed all personal relations. Drawing on Lewis Henry Morgan's model of early societies based on native American tribes, where all property was held communally and leaders were democratically chosen on their merits, she argued that women's power over familial and social relations had translated into their effective control over the tribe's means of subsistence. Her historical analysis parallels that of Friedrich Engels's *Origin of the Family, Private Property, and the State,* also based on Morgan's work, in significant ways. Postulating that women were the first tillers and cultivators of the soil, since men would have been too busy with warfare and the hunt, she stressed the importance of female labor and invention to feminine power and to civilized progress (*SSH* 150, 159–60), and traced woman's enslavement, along with the defeat of "liberty, equality, and fraternity, the cardinal virtues and principles of early society," to the triumph of the egotistical "excessive desire for property" under male rule (*SSH* 140). On the other hand, Engels considered communal marriage to signify human progress precisely because it represented a triumph over a natural animal jealousy and possessiveness. Given Gamble's condemnation of male sexual license, she would have joined other American socialists in advocating social purity and monogamy rather than Engels's vision of freer sexual unions as a goal of social reform.[10]

In the anthropological narratives, marriage by capture marked the first positive step in a history of family life that would eventually place women on their appropriately protected pedestals and make fathers' affection for their sons and heirs a reality, rather than representing, as Gamble believed, the defeat of the "progressive principle" (*SSH* 103) in human nature. Dismissing the possibility that primitive women had been capable of contributing to their own support or that of their tribe, McLennan argued that female infanticide would have been an obvious response in times of dearth. The resulting shortage of women had been corrected by polyandry in endogamous tribes and wife capture in exogamous ones (McLennan 58, 69–70). Lubbock rejected polyandry as too unnatural to be anything but exceptional and cast doubt on the prevalence of infanticide; rather, the natural desire to have sole possession of a woman, blocked by the proprietary rights of fellow tribesmen over communally held women, motivated men to steal wives from rival tribes (Lubbock 143, 153). Pride of ownership inspired affection and enforced chastity, which, by insuring paternity, made filial affection possible. For both scholars, the rise of private property cemented the transition to patriarchy by giving men an incentive to identify and cherish their male heirs.

Gamble's quite different assumptions about female nature and female power led to a distinctively different analysis of the sexual politics of cultural evolution. In her eyes, given women's natural sexual restraint and their behavior in uncorrupted savage societies, they were much more likely to have instigated the abridgment of conjugal rights than were males concerned with one another's proprietary claims (*SSH* 232). The fact that modern savage women did not "assert their rights," as Lubbock put it, was hardly surprising given their complete enslavement (*SSH* 229), but this was no bar to believing that matriarchy had been the earliest condition of human society. Gamble did credit Lubbock with implicitly realizing that marriage by capture had been motivated by the male desire for women without any claim to tribal rights or protection. As she put it, "the independence of free women was a bar to the gratification of the lower instincts in man" (*SSH* 236); captured wives had become powerless slaves not because of any inherent female inferiority or weakness, but because they lost all the privileges guaranteed by their matrilineal gens. Men profited doubly by being able to take control of the property abandoned by stolen sisters and wives; as wealth began to accumulate in male hands and hostilities with other tribes lessened, they voluntarily bartered their own freeborn women for powerless, foreign-born wives (*SSH* 192–93). In her view, female infanticide would never have been common practice, given women's real contributions to the primitive economy and the reality of maternal love, but rather the last act of desperation by mothers seeking to save their female children from capture (*SSH* 218–20). Similarly, Bachofen's Amazon revolt could be explained as an exceptional final stage of resistance by women trying to maintain their dignity and honor against the encroachments of male control (*SSH* 209).

Gamble went on to construct a history of marriage in which the progressive loss of economic power, rather than the triumph of superior male morality and intelligence, accounted for women's inferior position throughout much of history. Woman's historically greater capacity for sexual self-restraint proved to her that "the innate tendencies in the female constitution," not male tuition and training as Lubbock implied, had made civilization possible by checking our animal nature (*SSH* 237). Although woman's "passionlessness" often served evangelical ends in the nineteenth century (Cott), Gamble stressed instead the ways religion had played a dominant role in enforcing women's inferiority, whether in the replacement of goddesses by male deities or in the Pauline misogyny that contradicted Jesus's own teaching by enshrining a dogma of female depravity at the center of Christian faith (*SSH* 358–63; see Gamble's *The God-Idea of the Ancients* for a fuller development of this critique). For Gamble, it was woman's gradual decline into sexual slavery that had thrown her true moral superiority into eclipse. It was no wonder that modern women seemed weak and ruled by passion, when they had been completely dependent upon men for thousands of years; freedom and independence were necessary for woman's innate moral refinement to thrive (*SSH* 238). No wonder she seemed nervous and hysterical, when her own sexual preferences had so long been overruled by man's (*SSH* 54, 60). And no wonder Darwin found so few intellectually distinguished women, when all paths to success had been blocked by oppressive laws and customs throughout western history. On this last count, at least, Gamble took comfort in the evidence of woman's successful competition with men that had accumulated in the years since Darwin had made his claims (*SSH* 79).

In prescribing greater economic and personal freedom as the antidote to women's subordination, Gamble in some respects anticipated the position of Charlotte Perkins Gilman, whose *Women and Economics* was published in 1898. Gilman, too, seized on the discontinuity in female roles between animal and human society in *The Descent of Man* and deplored the fact that marriage was the middle-class woman's sole economic option. But where Gilman prescribed gainful employment outside the home to correct an inferiority that became inevitable once their dependence on men exempted women from the pressure of natural selection (Gilman 60–63), Gamble was more concerned with deflecting "the abnormal development of the reproductive energies in the opposite sex" (*SSH* 80) that had dominated human development once men gained control of women and production. Since women had become the economic and sexual slaves of men, "no male biped has been too stupid, too ugly, or too vicious to take to himself a mate and perpetuate his imperfections" (*SSH* 82); only woman's economic independence could make marriage a freely chosen partnership and reproduction a voluntary and progressive extension of the race. Although Gamble in a sense echoed Darwin's call at the end of the *Descent* for the eugenic self-discipline of the middle classes (Darwin 918–19) by looking toward a future when only the most robust women would reproduce themselves with worthy men (*SSH* 399), she was not about to reenslave women to the exclusive demands of motherhood. Indeed, she was deeply suspicious of church-sanctioned calls upon women to become mothers in her own day, seeing their real source as capitalism's endless appetite for a large and cheap workforce (*SSH* 392–93). She viewed the declining birthrate in her own day not as an ominous sign of impending "race suicide" but as an indication of a salutary increase in women's control over their own sexuality.

Responses to Gamble's work when first published as *The Evolution of Woman* were mixed; judgments of the book's scientific merits seemed directly related to the degree of the reviewer's agreement with her conclusions. While finding plausible Gamble's reconstruction of the position of woman under matriarchy and in later historical periods, the *Popular Science Monthly* disputed her scientific evidence that the higher faculties were transmitted solely through the female and insisted that "paternal love is also a primary instinct" (275) in both animals and man. The *Critic* reviewer applauded Gamble's attempts to explode the "dogma" of woman's inferiority to man; whereas the *Popular Science Monthly* argued that the book should have been called "The Rise and Fall of Woman, with a Prophecy of Her Renaissance" (275), the *Critic* claimed that Gamble presented "not a harangue on women's rights, but a careful and scientific treatise" (21). Most complimentary was the review in the *Nation*, written by Christine Ladd Franklin, herself a strong supporter of women's education and suffrage.[11] Although she implied that Gamble may have exaggerated the virtues of early women, Franklin basically endorsed her claims of the evolutionary benefits of female altruism and completely accepted her economic analysis of the position of women in matriarchal cultures (Franklin 452). Franklin recommended the book for general readers as well as for those interested in highly debated questions of social development (453).

Gamble's attempt to restore the absent woman in male accounts of progress and to rewrite the plot of civilization to foreground the struggles and eventual triumph of her biologically programmed virtues aimed, like the narratives she wrote against, to justify perceived truths about human nature by inventing a history for them. From Darwin's perspective, woman's inferiority to man seemed natural and required an evolutionary provenance; for the anthropologists, woman's exemption from labor and her subordination to man were the highest achievement on a single fixed scale of social development that they reconstructed. Gamble's vision of true womanhood has more in common with the logic of domestic ideology that Nancy Armstrong traces through the Victorian novel, in which female virtue overcomes male aggression and competitiveness and subordinates it to nurturing love. Yet if Gamble's handling of that virtue tends, like the novelists', to reduce social problems to differences of gender, she remains too aware of the economic causes of class difference to treat domesticity as the sign of class superiority as they did (Armstrong 4–8). Her radical essentializing of female altruism makes it the common possession of all women in a society free of competition, and she models human origins as this kind of socialist utopia. Nevertheless, embracing a biologically determined argument for woman's genius made Gamble's feminism vulnerable; Flavia Alaya points out that when nineteenth-century feminists exchanged an Enlightenment egalitarianism demanding that women too share in the rights of man for one that based its claims on woman's unique nature, they found the ground of innate difference an inherently unstable one; scientific justifications for complementary but equal gender roles could all too easily rationalize the same subordinate status quo and block real social progress (Alaya 260–63; see also Richards 96, 110n).

Our advancing scientific understanding has intensified, not silenced, the struggle for narrative hegemony among contending explanations for modern gender roles in the twentieth century. These narratives remain haunted by ghosts of the Victorian past. A version of the "traffic in women" little different from that devised by Victorian anthropologists provided the foundation for Lévi-Strauss's study of kinship relations and Freud's mapping of the family psyche (Rubin). The promiscuous Victorian primitive surfaced again in the constant state of female sexual receptivity that Haraway identifies as crucial to the extrapolation of the human family out of primate groups in early-twentieth-century primatology (23). The helpless female was essential to theories that made "Man the Hunter" the effective agent of the transition to full humanity, and the presence and power of the Father remained the determining fact of human nature and human society in physical anthropology (Haraway 220–21). Fuller investigation of the limits of sexual selection has done little to challenge or to explain the shift from female selection in animals to male selection in man (Zihlman 475–76).

Recent feminist counternarratives have contested some of these readings in ways that at times echo Gamble's positions. By stressing the role of woman in technological innovation, anthropological theories placed "Woman the Gatherer" back in the foreground of advancing civilization. Primatologists have redrawn the social life of key species like chimpanzees along matrifocal lines, and in the process of demonstrating the female's sexual prerogatives have in some ways credited her with an updated version of the Victorian power of female domestic virtue to shape male sociability to its own ends.[12] Nonetheless, we still echo Victorian debates in which choice means control and the division of labor justifies differential empowerment. At the end of the search for scientific justifications of cultural order, we still find only narrative, in which the relevance of biological "fact" remains inextricable from interpretation, analogy, and metaphor.

Eliza Burt Gamble (1841–1920): Major Works

The Evolution of Woman (1894)
The God-Idea of the Ancients (1897)
The Sexes in Science and History: An Inquiry into the Dogma of Woman's Inferiority to Man (1916) [revised and enlarged from *Evolution of Woman*]

Notes

1. See Donna Haraway's claim that explanation in the biological sciences is a storytelling practice in which rival reconstructions of the history of nature contend to justify desired present worlds (4–5).

2. Gillian Beer discusses the functions of such ordering truths in evolutionary allegory and analogy in *Darwin's Plots,* especially 80–90. James Clifford makes a similar point about the role of transcendent truths in "On Ethnographic Allegory," although noting that twentieth-century ethnographies have partially substituted "humanist allegories" for earlier historical ones (101–2).

3. For analyses of patriarchal biases in Victorian reconstructions of the history of the family, see in particular Fedigan, Fee, and Russett, chap. 3.

4. *The Sexes in Science and History,* from which my citations are taken, adds minor revisions to the first two-thirds of *The Evolution of Woman,* expands the discussion on women under Roman law to include a fuller account of the early Christian era, and adds a new chapter bringing the status of women up through the Renaissance and early modern period. It substitutes a new and slightly longer conclusion for the original.

5. See Tedesco for a summary and analysis of Blackwell's work. I take issue with her claim (65) that Gamble closely followed Blackwell's analysis. Russett also overgeneralizes in saying that like Blackwell, Gamble believed the sexes possessed "complementary and strictly equivalent strengths" (98).

6. Later commentators used the evidence of greater male variability to argue that men were overrepresented as both geniuses and idiots, while women occupied the mediocre middle in human achievement. See Russett's analysis of Havelock Ellis on this point (94).

7. Although she does not cite him, Gamble's reasoning here closely resembles some of Lester Ward's positions. Ward's 1888 article "Our Better Halves," while not challenging claims of woman's current physical and mental inferiority (268), described the female element as central to reproduction in the lower animals and considered woman "the unchanging trunk of the great genealogic tree" (275), while man was but a grafted branch. He advocated the elevation of woman's status as the only sure road to uplifting the race.

8. Armstrong (121–24) and Yeazell (chap. 12) elaborate on these parallels between animal and Victorian courtship.

9. See my "Darwin and the Anthropologists" for a fuller analysis of Darwin's rhetorical difficulties in the *Descent.*

10. Buhle (*Women and the American Left* 29) notes that although a translation of Engels's 1884 *Origin of the Family* was not published in the United States until 1902, most American socialists would have already been familiar with the outlines of its argument. Many supported social purity movements out of a desire to dissociate themselves from the popular perception that socialism advocated a free love promiscuity (Buhle, *Women and American Socialism* 250–52). Gamble's claims for primitive egalitarianism and female power go beyond Engels's and Morgan's positions. Although the conclusion to *The Evolution of Woman* stresses the historical exploitation of the poor by the wealthy and other abuses of capital-

ism (343), Gamble does not explicitly mention socialism until the 1916 edition (*SSH* 390–91), and there is no evidence to confirm the extent of her involvement, if any, in organized socialist movements.

11. Although the review is anonymous, *Poole's Index* credits Franklin as the author.

12. See Haraway's chapters on Hrdy and Zihlman for examples of these feminist approaches and of the androcentric prejudices they countered. She notes that Zihlman's studies of chimpanzees hypothesized that females would encourage the development of sociability by choosing the less aggressive and more socially skilled males as mates (Haraway 338). See also Fedigan's comprehensive summary of woman's role in theories of human evolution from the nineteenth century to the present.

Works Consulted

Alaya, Flavia. "Victorian Science and the 'Genius' of Woman." *Journal of the History of Ideas* 38 (1977): 261–80.

Armstrong, Nancy. *Desire and Domestic Fiction: A Political History of the Novel.* New York: Oxford UP, 1987.

Bachofen, Johann Jakob. *Myth, Religion and Mother-Right.* Trans. R. Mannheim. Princeton: Princeton UP, 1976.

Beer, Gillian. *Darwin's Plots: Evolutionary Narrative in Darwin, George Eliot, and Nineteenth-Century Fiction.* London: ARK, 1983.

Blackwell, Antoinette Brown. *The Sexes throughout Nature.* New York: Putnam's, 1875.

Buhle, Mari Jo. *Women and American Socialism, 1870–1920.* Urbana: U of Illinois P, 1981.

Buhle, Mari Jo. *Women and the American Left: A Guide to Sources.* Boston: G. K. Hall, 1983.

Clifford, James. "On Ethnographic Allegory." In *Writing Culture: The Poetics and Politics of Ethnography,* ed. James Clifford and George E. Marcus. Berkeley: U of California P, 1986. 98–121.

Cott, Nancy. "Passionlessness: An Interpretation of Victorian Sexual Ideology, 1790–1850." *Signs* 4 (1978): 219–36.

Critic 14 July 1894: 21 [Rev. of Gamble's *Evolution of Woman*].

Darwin, Charles R. *The Origin of Species by Means of Natural Selection, or the Preservation of Favored Races in the Struggle for Life and The Descent of Man and Selection in Relation to Sex.* New York: Modern Library, 1936.

Detroit Free Press. 19 Sept. 1920. 1:6 [Gamble Obituary].

"Eliza Burt Gamble." *National Cyclopedia of American Biography.* New York: James T. White, 1922. 18:220–21.

Engels, Friedrich. *The Origin of the Family, Private Property, and the State, in Light of the Researches of Lewis H. Morgan.* New York: International Publishers, 1942.

Fedigan, Linda Marie. "The Changing Role of Women in Models of Human Evolution." *Annual Review of Anthropology* 15 (1986): 25–66.

Fee, Elizabeth. "The Sexual Politics of Victorian Social Anthropology." In *Clio's Consciousness Raised: New Perspectives on the History of Women,* ed. Mary S. Hartman and Lois Banner. New York: Harper Colophon, 1974. 86–118.

[Franklin, Christine Ladd.] "Sex Predominance in Historical Development." *Nation* 58 (1894): 452–53.

Gamble, Eliza Burt. *The Evolution of Woman.* New York: Putnam's, 1894.

Gamble, Eliza Burt. *The God-Idea of the Ancients.* New York: Putnam's, 1897.

Gamble, Eliza Burt. *The Sexes in Science and History: An Inquiry into the Dogma of Woman's Inferiority to Man.* New York: Putnam's, 1916.

Geddes, Patrick, and J. Arthur Thomson. *The Evolution of Sex.* London: Walter Scott, 1889.

Gilman, Charlotte Perkins. *Women and Economics: A Study of the Economic Relations between Men and Women as a Factor in Social Evolution.* Ed. Carl N. Degler. New York: Harper Torchbook, 1966.

Haraway, Donna. *Primate Visions: Gender, Race, and Nature in the World of Modern Science.* New York: Routledge, 1989.

Hersh, Blanche Glassman. "The 'True Woman' and the 'New Woman' in Nineteenth-Century America: Feminist-Abolitionists and a New Concept of True Womanhood." In *Woman's Being, Woman's Place: Female Identity and Vocation in American History,* ed. Mary Kelly. Boston: G. K. Hall, 1979. 271–82.

Jann, Rosemary. "Darwin and the Anthropologists: Sexual Selection and Its Discontents." *Victorian Studies* 37, no. 2 (Winter 1994): 287–306.

Lubbock, Sir John. *The Origin of Civilisation and the Primitive Condition of Man.* 6th ed. New York: Longmans, Green, 1902.

McLennan, John. *Primitive Marriage.* Chicago: U of Chicago P, 1970.

Popular Science Monthly 46 (1894): 275 [Rev. of Gamble's *Evolution of Woman*].

Richards, Evelleen. "Darwin and the Descent of Women." In *The Wider Domain of Evolutionary Thought,* ed. David Oldroyd and Ian Langham. Dordrecht: Reidel, 1983. 57–111.

Rubin, Gayle. "The Traffic in Women: Notes on the "Political Economy" of Sex." In *Toward an Anthropology of Women,* ed. Rayna Reiter. New York: Monthly Review P, 1975. 157–210.

Russett, Cynthia Eagle. *Sexual Science: The Victorian Construction of Womanhood.* Cambridge: Harvard UP, 1989.

Stocking, George W., Jr. *Victorian Anthropology.* New York: Free P, 1987.

Tedesco, Marie. "A Feminist Challenge to Darwinism: Antoinette L. B. Blackwell on the Relations of the Sexes in Nature and Society." In *Feminist Visions: Toward a Transformation of the Liberal Arts Curriculum,* ed. Diana L. Fowlker and Charlotte S. McClure. University: U of Alabama P, 1984. 53–65.

Ward, Lester. "Our Better Halves." *Forum* 8 (1888): 266–75.

Yeazell, Ruth. *Fictions of Modesty: Women and Courtship in the English Novel.* Chicago: U of Chicago P, 1991.

Zihlman, Adrienne. Review of *Sexual Selection and the Descent of Man,* ed. Bernard Campbell. *American Anthropologist* 76 (1974): 475–78.

9

Revisioning Darwin with Sympathy
Arabella Buckley

Barbara T. Gates

To set the stage for the entry of Arabella Buckley and her work in popularizing Darwin into this volume, I would like to go back nearly a hundred years before her late Victorian time—to the turn of the preceding century. Then, and continuing on into the Victorian era, women were called upon to aid children in developing respect and sympathy for nonhuman species. They were reinforced in this role by such influential figures as Erasmus Darwin, who in his *Plan for the Conduct of Female Education in Boarding Schools* (1797) recommended that women inculcate in other young women "sympathy, when any thing cruel presents itself; as in the destruction of an insect" (47). Darwin went on to illustrate his point: "I once observed a lady with apparent expressions of sympathy say to her little daughter, who was pulling off the legs of a fly, 'how should you like to have your arms and legs pull'd off? would it not give you great pain? pray let it fly away out of the window': which I doubt not would make an indelible impression on the child, and lay the foundation of an amiable character" (47). This particular kind of moral responsibility would be repeatedly reinforced in conduct books like Maria Grey and Emily Shirreff's *Thoughts on Self-Culture, Addressed to Women* (1851), where women again were charged with the duty of inculcating in the young sympathy and benevolence for all "inferior beings." Still later in the nineteenth century, Darwin's grandson,

Charles, would influence Patrick Geddes and J. Arthur Thomson in their widely read study *The Evolution of Sex* (1889), which characterized women as having evolved "a larger and habitual share of the altruistic emotions," and as excelling "in constancy of affection and sympathy" (270–71).

Small wonder, then, that a need and desire to sympathize remained with women even when the new generation of post-Darwinian women appeared toward the end of the nineteenth century. Remarkable among these was Arabella Buckley, Sir Charles Lyell's secretary. Personally familiar with the leading scientists and scientific theories of her day, Buckley was a knowledgeable and authoritative popularizer of science who also accepted a woman's responsibility to both educate and teach morality to the uneducated and the young. In her first book, *A Short History of Natural Science* (1876), she recalled with a firm voice how she "often felt very forcibly how many important facts and generalizations of science, which are of great value in the formation of character and in giving a true estimate of life and its conditions, are totally unknown to the majority of otherwise well-educated persons" (vii–viii). This is to say nothing of children, who are fed only a few elementary and scattered facts about science, and therefore are unprepared "to follow intelligently the great movement of thought" (viii). Buckley's own book was intended "to supply that modest amount of scientific information which everyone ought to possess, while, at the same time . . . form a useful groundwork for those who wish afterwards to study any special branch of science" (viii), and as such was praised by Charles Darwin.

History obviously fascinated Buckley, but it never gave full scope to her distinctive penchant for narrative. Only the story of evolution did that. Thoroughly grounded in evolutionary theory and in all aspects of the new geology, Buckley set out this knowledge in a series of books whose narratives are brilliantly original. It is difficult to imagine a popularizer of Darwin who could outnarrate Darwin's own remarkable story, with its high drama of struggle and its presiding Nature who both is and is not the principle of natural selection. But, as Darwin himself knew, the story of evolution was many stories. His revisions and his own later comments on what he meant by natural selection show his own openness to reinterpretation. Like Huxley, Buckley was one of many later interpreters of Darwin who took advantage of the capaciousness of the story of evolution.

This she did in two highly imaginative volumes, *Life and Her Children* (1881) and *Winners in Life's Race* (1883). In these books, Buckley presented seven divisions of life: *Life and Her Children* covers the first six, from the amoebas to the insects, and *Winners in Life's Race,* the seventh, the "great backboned family." Life, or the life force, is only a partial heroine in this story. It drives creatures to continue life for their species but does not exhibit the sympathy that Buckley would most admire in a heroine. That role would be reserved for the vertebrates of the animal kingdom. In this way, as in many ways, Buckley's story does not recapitulate Darwin's, with its confusion between a benevolent-seeming natural force, Nature, and a more indifferent one, natural selection.

In *Life and Her Children,* the first of the two volumes on evolution, Buckley reviewed the struggle for existence and adaptations of the simpler animals, concluding with an interesting appraisal of ants that prepared for her second volume. Her entire chapter on ants is highly authoritative: Buckley supports every observation and anecdote with scientific accounts by eminent authorities and leads us to consider ant colonies throughout the globe. But ants left her with a final troubling question: whether these creatures, with all their socialized behavior, were also marked by sympathy. Her conclusion was that the ant's place in Life's scheme need allow for no such thing—ants are devoted to community, not to each other. It remained for the backboned creatures to illustrate kindness.

Winners shows these backboned creatures in far greater detail than its predecessor had depicted the invertebrates and reveals the original ways in which Buckley handled narrative problems that had also faced Darwin. Having written a history in her first work, Buckley knew well how to depict events imagined through time. But the story of evolution called for a different kind of history, a different kind of time. The need to get beyond lifespans, not just of individuals but of entire species, forced Buckley to reach past the strategies she had previously relied upon. The need to describe a history that was preverbal and antecedent to human consciousness sent her deep into images (figures 9.1 and 9.2).

In an early chapter Buckley recalls how "with a history so strange" she wishes to "open the great book of Nature still further, and by ransacking the crust of the earth in all countries . . . try and find the explanation, which will no doubt come some day to patient explorers" (210)—the explanation of the "missing links." What stops Buckley in her long and colorful recounting of the evolution of species, told as spectacle and with a panoramic vividness that paints pictures rather like museum dioramas, is a "strange blank"—the mysterious end of the age of reptiles. This sends her back to a "history" (342) of the mammals and humankind's appearance on the scene, which in turn allows

Figure 9.1. An image of early mammals. Illustration from chapter 9, *Winners in Life's Race.*

her a place for her favorite improvement upon Darwin: her belief in the evolution of sympathy.

Before she launches into her explanation of this phenomenon, Buckley looks for awhile at adaptation, then instinct. These usher her into the realm of feeling, exactly where she wishes to be. The loyalty of pet snakes, the sacrificial miming of injuries by parent birds trying to distract enemies away from their young, the instinct of herding for protection—all these lead her to speculate that "one of the laws of life which is as strong, if not stronger, than the law of force and selfishness, *is that of mutual help and dependence*" (351). This law was not a special gift to human beings, as Christians might like to believe, but a gradual development through the animal world. In considerable contrast to the social Darwinians of her day, Buckley concludes her book with "the great moral lesson taught at every step in the history of the development of the animal world, that amidst toil and suffering, struggle and

Figure 9.2. Humankind and other hunters emerge after the Ice Age. Illustration from chapter 12, *Winners in Life's Race.*

death, the supreme law of life is the law of SELF-DEVOTION AND LOVE" (353). For Buckley, the *raison d'être* for evolution was not just the preservation of life, but the development of mutuality as well.

Buckley's last book, *Moral Teachings of Science* (1891), was devoted to this idea and written to unite science and philosophy—to study morality from "within outward" and "without inward" (4). For Buckley, "these are not really two, but only different methods of arriving at one result, namely, the knowledge of laws by which we and all the rest of nature are governed" (5). Life for her had become universal; she set out to examine a natural and human world where struggle predominates but mutality works to benefit individuals of all groups, including human beings. Although Buckley plants us firmly in the world of sympathy, as a post-Darwinian she takes us entirely out of the pre-Darwinian moral order of William Paley (*Natural Theology,* 1802), who had tried to find evidences of God in nature. "We must remember," she reminds us, "that this [evolution, development] has not taken place by special guidance along certain beneficent lines, since degradation and partial deformities result as by-products of the struggle for life; but that the overwhelming preponderance of healthy, happy, and varied existence has been brought about by the steady working of natural laws among which the struggle to survive and the constant action of natural selection are the most important" (54).

In this championing of Darwin, as in her belief in mutuality, Buckley looked more to the future than to the past. One could, of course, view Buckley's deep concern with cooperation as distinctly Victorian sentimentality or only as a part of the female tradition of the teaching of sympathy that I have described for you, and could minimize its significance on both grounds. In recovering this important woman popularizer I prefer, however, to leap ahead to see in Buckley's vivid and sympathetic work something more visionary, and also more original—a significant rewriting of Darwin for the future. Buckley was one of a small number of the early Darwinians who realized the deficiencies in Darwin's thinking with regard to the development of moral qualities in the animal kingdom, set out in his discussion of "social instincts" in the *Descent of Man* (1871). Darwin had observed the competitive advantage derived from well-developed social instinct but was at a loss to explain it, particularly with respect to parental affections. Far from being daunted by this aspect of evolution, Buckley, with her title *Life and Her Children,* made parenting her central metaphor and continued Darwin's observations with far greater emphasis on mutuality.

In this she was a pioneer. Her work is concurrent with Karl Kessler's "On the Law of Mutual Aid" (1880), the lecture that stimulated Peter Kropotkin to reexamine Darwin. Kessler died in 1881, the year that saw the publication of Buckley's *Life and Her Children;* it then took Kropotkin ten years to challenge Thomas Henry Huxley over the importance of mutual aid in the pages of *Nineteenth Century,* and another twenty to formulate his classic *Mutual Aid: A Factor in Evolution* (1902). Huxley, you will recall, had claimed that "cosmic nature is no school of virtue, but the headquarters of the enemy of ethical nature" (75). Had Buckley been writing for a different audience, her work might have entered the debate, have helped counter Huxley and have filled an intellectual lacuna that, on account of Kropotkin's hesitations, remained for over twenty years.

To move briefly into another sphere and look ahead still further, Buckley was even more in advance of her time in discussing the life force than she was in advocating mutuality as an evolutionary principle. Buckley's "Life" has much in common with the Gaia of the hypothesis only recently set forth by Lynn Margulis and James Lovelock. For them, Gaia, like Buckley's "Life," becomes an appropriate metaphor to describe that collectively evolving organism in whose evolution cooperation and mutalism have been even stronger forces than has competition.

Buckley was not only fully aware of new directions in science; she was equally aware of the nature of science writing itself—of both its pitfalls and its possibilities. She realized that science, though based in fact or experiment, was transmitted as a literary construction. The acceptance of science depended upon effective persuasion, including skilled use of narrative and metaphor. To further illustrate the metaphoric aspect of Buckley's work, I will briefly discuss two other books, *The Fairy-Land of Science* (1879, reissued in a number of late-nineteenth-century editions) and its sequel, *Through Magic Glasses* (1890). In these books, Buckley generated interest in her scientific subjects by borrowing the language of fairy stories and wizardry to reinforce her ultimate belief that the wonders of science not only paralleled but surpassed the wonders of fairyland (figure 9.3). The *Fairy-Land of Science* is introduced with an epigraph from folklore, suggesting that fairy tales imprint the memory in ways that Buckley hopes science will. After briefly telling the story of Sleeping Beauty, on her second page Buckley asks: "Can science bring any tale to match this?" Her answer is a resounding "yes"; her fairies, which are "forces" like magnetism and gravity, will be every bit as fascinating as Aladdin's genie or Ariel and Puck. In her sequel, *Through Magic Glasses,* Buckley would focus more closely on what childlike eyes can see, here with the help of the telescope, stereoscope, photo camera, microscope, and a fictional guide, a magician whose chamber—and eyes—we enter with the first pages of the book.

What I believe Buckley was doing in both of these books was something characteristic of post-Darwinian scientific writing: attempting somehow to transcend the limitations of the human eye and human realistic language as adequate focalizers and descriptors of natural phenomena. In several ways children's books left Buckley freer to do this than Darwin's chosen media—the treatise and scientific paper—had left Darwin. As Gillian Beer and James Krasner have both told us, Darwin's biology seems to have made his writing more difficult for him. When Darwin chose to tell the story of evolution, he realized that the tale of life and its origins and development needed an omniscient narrator but, given the evolutionary shortsightedness of *homo sapiens,* could not really have one. He must settle for describing nature through the inadequate lens of the human eye and with the inadequate verbal constructs of a realistic literary tradition. Hence his careful decodings of

THE

FAIRY-LAND OF SCIENCE.

LECTURE I.

HOW TO ENTER IT; HOW TO USE IT; AND HOW TO ENJOY IT.

I HAVE promised to introduce you to-day to the fairy-land of science,—a somewhat bold promise, seeing that most of you probably look upon science as a bundle of dry facts, while fairy-land is all that is beautiful, and full of

B

Figure 9.3. First page of *The Fairy-Land of Science,* 1892 edition.

metaphorical tangled banks, brilliantly attempting to compensate for omniscience through the proliferation of written detail. After Darwin, the imperfect eye of the physical human organism, suggesting as it did the interpreter's own evolutionary deficiencies, became a bane of science writing.

Working in the area of scientific popularization, Arabella Buckley could, however, write science with a difference. Transgressing the borderlands of acceptable scientific forms, Buckley defied the limitations of both realistic language and the human sensory experience. In *Fairy-Land,* she could comfortably use the language of illusion to describe the invisible world of forces, and then in *Magic Glasses* she could reinforce the importance of access to humanly designed machines that help correct for flawed human vision.

Before going deeper into these two books, I would like to backtrack a moment to look briefly at the writing of Sir Charles Lyell. Like Darwin's, Lyell's work was highly visual, but its focus was not the close-up. Lyell liked to envision the scope of things, to imagine vast panoramas or deep cross-sections that sent his reader's eye back through time or downward into the unseeable earth's crust. In doing so he did not choose to limit himself to the human eye, but asked his readers to imagine looking through the eyes of other creatures, to try to see the oceans, for example, through those of an imaginary amphibian with human powers of understanding. Then, according to Lyell, one might "ascertain, by direct observation, the action of a mountain torrent, as well as of a marine current; might compare the products of volcanos with those poured out beneath the waters; and might mark, on the one hand, the growth of the forest, and on the other that of the coral reef" (1:99). Such vivid, persuasive, and compensatory narrative strategizing was well-known to Buckley, who worked closely with Lyell from 1864 to 1875, taking dictation for the geologist, copying his texts, and no doubt discussing his writing as they worked alongside one another. As we have seen, Buckley herself utilized similar narrative techniques in *Life and Her Children* and *Winners in Life's Race.*

Then, in *Fairy-Land,* Buckley went further. If Darwin's discourse was marked, as Krasner believes it was, by a tension over "misprision, illusion, and limitation" (5) that led him and other scientists to an empirical self-consciousness, Buckley's exploded the worry and self-importance implicit in such self-consciousness. Writing on the periphery of established science, Buckley came to envision—and even to revel in—nature writing as a more accurate and correctable form of fiction and told her readers so at every turn. If such writing was grounded in empirical evidence, it was nevertheless clearly a mental construction, based in inference as well as in observation. Children reading her science writing might find it both more wonderful and more exacting in describing what they did and did not know than were most fairy stories, but they were nevertheless told it was similar and approximate. According to Buckley, "all this which is true of the fairies of our childhood is true too of the fairies of science" (6).

Why, then, should children or anyone else bother to read science at all? The reasons lay in its greater truth and in its utility. Take for instance Buckley's own example: the story of a knight who attempts to cross a raging torrent and just before being swept away is saved by a water nymph who guides him to the opposite shore. A person could be like this knight and attempt to dash across a torrent, hoping for rescue by a supernatural water nymph. Or a person could be informed by a scientific principle that would allow him or her to read the currents and make for the shore in due course and due time. Both water nymph and current reading are learned constructions, but the one is more likely to save one's life than the other. Buckley shows her readers how to grow from one set of beliefs to another, from water nymphs to trigonometry and liquid flow, telling them what equipment they will need to make the journey with her.

First of all, they must open their eyes; they must learn to look at what is around them in the universe. But then they must also learn to exercise imagination, for much is "hidden in the things around them" (913). Buckley's young niece once wanted to know why there was a mist on her bedroom window in the morning. Buckley then breathed on the window pane, showed her the mist, and reminded her that "Cissie and auntie have done this all night in the room" (14). According to Buckley, this is how scientific knowledge can best be assimilated; forces can be understood only if one first discerns them at work and then questions the mystery that they represent. Finally, Buckley's readers must learn something of the language of science. "Not hard scientific names, for the best books have the fewest of these" (15), says Buckley, but simpler words, like "liquid" and "solid," so that they may enter the land of science like a well-equipped traveler, knowing its tongue.

Here of course language and the eye are foregrounded once again, but strongly mediated by imagination, whose powers Buckley pushes to the forefront. Sounding like a latter-day Coleridge, Buckley is careful to warn that she means for children studying science to exercise not "mere fancy, which creates unreal images and impossible monsters, but imagination, the power of making pictures or *images* in our mind, of that which *is*, though it is invisible to us" (7). This suggests an important distinction between fairy stories and science stories. Buckley will impart knowledge that will allow the young to see the invisible within the visible, not by visualizing a gothic being such as Frankenstein's Monster, but by apprehending contemporary and revisable scientific theory which is also based in human imagination. If such theory is a construct, a fiction, its probability and revisability bring it closer to describing what we see with our own two eyes. The human mind thus corrects for the human eye, but the two function together in shaping a view of the world.

Although Buckley's rhetoric and written medium allowed her a greater degree of freedom from empirical self-consciousness than scientific papers and scientific treatises might have allowed, Buckley still faced the problem of presenting the limitations of the human eye as an instrument of discovery. Here again, her medium was a help. Women popularizers had been writing primers about microscopes and telescopes for over a century, hoping to bridge a gulf between what Barbara Stafford calls mere " 'curious' *watching* and a rational, tasking, language-driven *observation*" (96). In Buckley's own time, Mrs. Ward had produced popular works on the telescope and microscope, hoping in the second of these to "attract those readers who, unversed in microscopic marvels, might possibly feel repelled by a complete and lengthened treatise" (v–vi). Clearly, her aim was to promote observation; her chapters set out to describe what one could see with the aid of a microscope. Along the way, Ward also felt obliged to describe the workings of the human eye peering at the eyes of other animals through the enhanced image of the microscope. Rather than imagining life through other eyes themselves as had Lyell, Ward advocated objectifying the vision of other species in order to understand it.

If Ward's was a standard way of popularizing optics, Buckley again tried something else. Drawing in readers by setting up a mysterious laboratory inhabited only by moonlight and a magician, she begins *Through Magic Glasses* with illusion. Before long, however, she demystifies her magician. He is in fact a knowledgeable principal of a school for boys of the artisan class—a man of science. Here Buckley seems only to have snared us with her talk of magic. The metaphor has been pure analogy; we are ushered into a lecture hall and are readied to learn optics. All human beings, we now learn, can sharpen their sight and become true magicians (read scientists). Like so many popularized versions of scientific discovery, *Through Magic Glasses* then asks its young readers to contribute to scientific inquiry. In her magician's words to his boys, "the value of the spells you can work with my magic glasses depends entirely upon whether you work patiently, accurately, and honestly." If they do so, the magician's students can then "look deep below the outward surface of life . . . and help to pave the way to such grand discoveries as those of Newton in astronomy, Bunsen and Kirchhoff in spectrum analysis, and Darwin in the world of life" (54). The magic of science is not arcane and black, Buckley seems to be saying, but accessible and white. Simple language and simple analogies can make science available to everyone and help correct for ignorance of the universe, just as improved optical devices can make it possible better to visualize that universe. "In these days, when moderate-priced instruments and good books and lectures are so easily accessible," says Buckley in her preface, "I hope some eager minds may be thus led to take up one of the branches of science opened out to us by magic glasses; while those who go no further will at least understand something of the hitherto unseen world which is now being studied by their help" (v). In either case, word and eye will facilitate each other.

Still, scientific mysteries will remain and give Buckley room to develop her own metaphors. As we have seen, Buckley's is a science of subjects, not objects. And so *Through Magic Glasses* concludes not with what we might see with optical devices, as in Mrs. Ward's book, but with what we might *not* see—the visionary and the imaginative, the literary mediation between the eye and scientific fact. In its final chapter, Buckley remystifies the old magician who sits alone in his study, concerned about how to present the story of human evolution. Like Darwin, Lyell, and Buckley in her earlier books on evolution, he must fall back upon "a waking dream" (210) based upon partial knowledge of a scattered fossil record. Puzzled, his students do not know why he seems possessed, with a "far-away look in his eyes" (226)—those same eyes that they had seen so effectively at work with optical devices enabling better sight. The boys strain to understand that he has "passed through a vision of countless ages" (226) and returned to tell them a yet incomplete story. Here Buckley references the self-consciousness of contemporary science, aware of itself as nature trying to decode nature with inadequate sensory and intellectual capabilities. But Buckley seems also to have realized that these very weaknesses in perception and lapses in the story of science offered her her main entry points as storyteller. If at the turn of the century Buckley and other women who were scientifically educated were not yet culturally deputized to carry the burden of scientific discovery, scientific popularization gave them elbowroom to explore the fictiveness of science writing.

Arabella Buckley (1840–1929): Major Works

A Short History of Natural Science (1876)
The Fairy-Land of Science (1879)
Life and Her Children (1881)

Winners in Life's Race (1883)
Through Magic Glasses (1890)
Moral Teachings of Science (1891)

Works Consulted

Beer, Gillian. *Darwin's Plots: Evolutionary Narrative in Darwin, George Eliot, and Nineteenth-Century Fiction.* London: Routledge and Kegan Paul, 1983.
Buckley, Arabella. *The Fairy-Land of Science.* London: Edward Stanford, 1879.
Buckley, Arabella. *Life and Her Children.* London: Edward Stanford, 1881.
Buckley, Arabella. *Moral Teachings of Science.* London: Edward Stanford, 1891.
Buckley, Arabella. *A Short History of Natural Science.* London: John Murray, 1876.
Buckley, Arabella. *Through Magic Glasses.* New York: D. Appleton, 1890.
Buckley, Arabella. *Winners in Life's Race.* London: Edward Stanford, 1883.
Darwin, Charles. *The Descent of Man, and Selection in Relation to Sex.* London: John Murray, 1871.
Darwin, Erasmus. *Plan for the Conduct of Female Education in Boarding Schools* (1797). New York: S.R., 1968.
Geddes, Patrick, and J. Arthur Thomson. *The Evolution of Sex.* London: Contemporary Science Series, 1889.
Huxley, Thomas H. *Evolution and Ethics.* New York: D. Appleton, 1894.
Krasner, James. *The Entangled Eye: Visual Perception and the Representation of Nature in Post-Darwinian Narrative.* Oxford: Oxford UP, 1992.
Kropotkin, Peter. *Mutual Aid: A Factor in Evolution.* London: William Heinemann, 1902.
Lovelock, James. *The Ages of Gaia: A Biography of Our Living Earth.* New York: Oxford UP, 1988.
Lyell, Charles. *Principles of Geology.* 11th ed. New York: Appleton, 1887.
Marcet, Jane. *Conversations on Chemistry, Intended More Especially for the Female Sex.* London: Longman, Hurst, Rees and Orme, 1805.
Paley, William. *Natural Theology; or, Evidences of the Existence and Attributes of the Deity, Collected from the Appearances of Nature.* London: Faulder, 1802.
Stafford, Barbara. "Voyeur or Observer? Enlightenment Thoughts on the Dilemmas of Display." *Configurations* 1, no. 1 (1993): 95–128.
Trimmer, Sarah. *Fabulous Histories, Designed for the Instruction of Children, Respecting Their Treatment of Animals.* London: Longman, Robinson, and Johnson, 1789.
Ward, Hon. Mrs. *The Microscope.* London: Groombridge and Sons, 1870.
Wollstonecraft, Mary. *Original Stories from Real Life.* London: J. Johnson, 1788.

PART 6

DEFINING AND REDEFINING KNOWLEDGE: INTO THE TWENTIETH CENTURY

10

Conflicting Scientific Feminisms

Charlotte Haldane and Naomi Mitchison

Susan Squier

> Scientific discourse served women as a vehicle for conveying arguments about female cultural politics. (Benjamin 40)

> Practicing conflict is also practicing feminism. (Childers and hooks 70)

> For me the unity of women is best understood not as a *given*, on the basis of a natural/psychological commonality; it is something that has to be worked for, struggled towards—*in history*. (Mohanty 84)

Contemporary feminist theory has sensitized us to the fact that debates about female cultural politics take place not only across the gender divide, but between women as well.[1] I want to use that insight to extend our analysis of women writing about science, using it to move beyond the crucial reclamation of women's popular science writing as an arena in which women have demonstrated their involvement with scientific practice and exerted power and agency to consideration of the issues and positions being advanced and debated when women write about science. My concern in this essay is to tease out the differences—indeed, the conflicts—between feminist positions. I will do so through an analysis of the writings of two modernist women science writers: Charlotte Burghes Haldane and Naomi Haldane Mitchison.

Like many female popularizers of science, Naomi Haldane Mitchison was born into a scientific family; like others, Charlotte Burghes married into one. Mitchison was the daughter of Oxford physiologist John Scott Haldane, and the younger sister of J. B. S. Haldane, the geneticist and popular science writer. With J. B. S.'s marriage to Charlotte Burghes, the two women became sisters-in-law for a time. In addition to this relation to Jack Haldane, Naomi Mitchison and Charlotte

Figure 10.1. Charlotte Haldane. Photograph from *Woman Today* (London: August, 1939).

Haldane shared a fascination with scientific practice and a determination to communicate scientific findings to the general public. Yet despite these commonalities, they were far from united in their views on or ways of writing about science. A comparison of the scientific writings of these two women, both of whom explicitly defined themselves as feminists and yet whose feminisms were dramatically different in content and method may expand our understanding of the tensions and possibilities at play when modernist women writers turned to scientific discourse to articulate feminist issues.

Although both Haldane and Mitchison were drawn to popular science writing as an authoritative site on which to stage arguments for women's increased agency, voice, and social position, the uses they made of scientific discourse were as different as their feminism and their class origins. They used scientific discourse to authorize two quite different arguments for woman's agency as a reproductive/sexual body, Charlotte Haldane in *Motherhood and Its Enemies* (1927) and Naomi Mitchison in *Comments on Birth Control* (1930). In their science writings they exemplify two conflicting feminist stances toward science: a valorization of the maternal role accompanied by an affirmation

Figure 10.2. Naomi Mitchison. From *An Outline for Boys and Girls and Their Parents,* edited by Naomi Mitchison (London: Victor Gollancz Ltd., 1932), illustrated by William Kermode and Ista Brouncker.

of normal science, and alternatively an appreciation of the multiplicity of women's biological and social possibilities accompanied by a revisionary drive to expand our notions of both scientific practice and scientists. As a brief survey of some of their popular science work will demonstrate, the different positions on science the two women held reflected their conflicting feminist commitments. Haldane's pronatalist, essentialist feminism led her to scientific meliorism, while Mitchison's proto-postmodern feminism led her to challenge the very disciplinary and epistemological premises of scientific practice.[2]

The horizon of expectations against which we read the popular science writings of Naomi Haldane Mitchison and Charlotte Haldane was set by J. B. S. Haldane in his classic essay, "How to Write a Popular Scientific Article." While he came to fame as a popularizer of science with his *Daedalus, or Science and the Future* (1923), an essay that predicted the revolutionary changes in domestic, social, and political life that would result from the application of science to human reproduction, it was with the later article that J. B. S. Haldane codified the accomplishment of his youth. "Most scientific workers desire to spread a knowledge of their subject and to increase their own incomes," he magisterially begins his advice for popular science writers (3). Yet a gender bias narrows Haldane's horizon in this essay: if the scientific worker is a woman, her scientific activity has already been curtailed by her gender. She has often been relegated to the margins, as craftworker, folk healer, teacher, or what Londa Schiebinger has called "invisible assistant" to male scientists (8). Moreover, when the science writer is a woman she may have motivations unanticipated by J. B. S. Haldane. Rather than hoping to spread a knowledge of her subject and increase her income, she may simply wish to achieve a foothold or to attain authority in science.

She may also hope to use science to authorize a feminist position. "Apart from my early acquaintance with the menace of anti-semitism, the second greatest influence on my youthful mind was that of feminism," Charlotte explained in her autobiography (*Truth Will Out* 6). Naomi Mitchison began, and then abandoned, a book on feminism in the 1930s. Later, she mused in her diary, "my feminism is deeper in me than, say, nationalism or socialism: it is more irrational, harder to argue about, nearer the hurting core." Although Mitchison rejected the label "feminist" in her old age, her writing and political activities demonstrate her lifelong, deep allegiances to feminism and socialism (Benton 63, 128). Yet in their era, to speak as feminists and achieve credibility was a difficult task. "By the late nineteenth century it was already necessary to demonstrate a scientific approach in order to gain full recognition," Jane Lewis has observed. "The ideas formulated by scientists and mediated by the medical profession formed the framework within which all women, including active feminists . . . had to work" (82).

Though as feminists they shared a strategic interest in science, Charlotte Haldane and Naomi Mitchison came to popular science writing by different routes, dramatizing a factor at times hidden by a feminist focus on gender and science: how class shapes a woman's development.

Naomi was educated at Lynam's School and Oxford University, where formal science training extended the science that was an inevitable part of her home life as the daughter of John Scott Haldane, reader in physiology at New College, Oxford. While only a teenager, she wrote a play about modern genetics that was given a full performance at Lynam's School, and "for a long time it was thought," she recalls in her memoirs, "that I would have made a good scientist" ("All Change Here" 63). Even after she had decided to abandon science in favor of writing, she had the background and the family connections that made it possible for her to write knowledgeably on a variety of scientific matters. In the jacket copy to her 1976 novel *Solution Three,* Mitchison describes herself as "coming from a family of scientists." Yet while Mitchison's family connection was clearly important, it would be an exaggeration to say that connection ensured automatic acceptance for her science writing. As she put it to me in a letter, "Scientists think I am frivolous, and non-scientists think I make things difficult."

In contrast, Charlotte Burghes Haldane's scientific education was limited to "the elementary bases of physics, chemistry and biology I had learnt at my German school in Belgium; barely enough to enable a schoolboy to pass matriculation" (*Truth Will Out* 16). Her plans for further education after her German secondary school were dashed by her father's business reversals. Forced to become a secretary, she only gradually found her way into journalism. When she decided to write a novel about science, she realized she required a scientific advisor. Unable to turn to any relative or close friend, as could Mitchison, she boldly sought out the author of the popular science essay that had impressed her with its vision of reproductive innovations, *Daedalus, or Science and the Future.* That man—who later became her husband—was J. B. S. Haldane.

Their class positions also shaped the positions on marriage and motherhood that Mitchison and Haldane turned to scientific discourse to articulate. Haldane, who described herself in her autobiography as a woman financially unable to stay home and care for her child, invoked scientific expertise to argue the case for subsidized motherhood. In contrast, Naomi Mitchison—whose work as a novelist was a matter of choice, not financial necessity, and who had nursemaids to help with the care of her five children—challenged the existing scientific technologies of contraception as giving a still-incomplete solution to woman's desire to combine employment, sexual freedom, and maternity. Note the divergence: both positions were designed to increase woman's agency, yet one conceptualized that agency as achieved through and focused on maternal involvement (so Haldane stressed increasing the social support for women as mothers), while the other saw it as expressed in a variety of venues, maternal, sexual, and occupational. So Mitchison argued that feminists will not want to be forced to choose between work and maternal/sexual love. They will want both worlds, and should be able to have both:

> Intelligent and truly feminist women want two things: they want to live as women, to have masses of children by the men they love and leisure to be tender and aware of both lovers and children: and they want to do their own work, whatever it may be. The two things are not compatible, except in very rare cases. . . . Adequate contraceptive methods are an essential part of this compromise. (*Comments* 25)

Both writers also constructed the social role of science differently. The duty of scientists, according to Charlotte Haldane, was to inform state policy, by delineating the difference between "normal" women who were qualified to become state-supported vocational mothers and abnormal, "intersexual" women—feminists and career women—who tended to disrupt social relations. "To give such females political and social power . . . may prove in the end the means to inactivate or to endanger those who must first of all be encouraged and protected," Haldane argued in *Motherhood and Its Enemies* (136). In contrast, again, Mitchison saw scientists as far less authoritative arbiters of social policy. In her *Comments on Birth Control,* she wrote of the contraceptive technologies they produced as "a compromise, and all compromises are by nature uninspiring and un-universal" (31). Yet she concluded that the difficult mood of sexual self-consciousness produced by the contraceptive compromise, while "a bad business on the whole, . . . is for the moment not only inevitable, but necessary," and she went on to acknowledge that "the whole problem of women's work, and especially married women's work, is in flux" (32).

While this divergence in the ways that Haldane and Mitchison treated the scientific power over women's bodies reveals the different agendas of pronatalist and sexual reform feminists, another moment of parallel popular science involvement by the two writers reveals a different sort of contrast; in the position they take toward science and the scientist. In the 1920s and early 1930s, both women took on a managerial or editorial role in popular science writing: Charlotte first as literary agent for J. B. S. Haldane and later as editor of the socialist women's magazine *Woman Today,* and Naomi as editor of the controversial *Outline for Boys and Girls and Their Parents* (1932). The assumption of such roles was a characteristic and important strategy of the female science writer, for they provided an autonomy and agency otherwise hard to come by in a profession implicitly (albeit unconsciously) gendered male. Such managerial/editorial roles in popular science writing merit a closer look by feminist scholars, for they represent another way that women have been able to "do science" (Eden 581).

Charlotte Haldane acknowledged this explicitly in her autobiography, recounting that it was she who gave J. B. S. the idea of writing popular science articles in order to capitalize on the success of his *Daedalus:*

> I foresaw the demand that would follow, for popular articles by him on scientific themes . . . he willingly accepted my suggestion that he should practice it [scientific journalism] as a hobby in his spare time, especially as it could be made to pay well. We decided that I would become his secretary and his agent. (*Truth Will Out* 21)

In later years, when she spoke with a BBC interviewer after Haldane's death, Charlotte expanded humorously on her role as J. B. S.'s secretary and agent, making it clear that she felt she had been responsible for a large part of his creative production. "I began to feed questions into this human computer," she recalled, "and out would come the explanations as to how things and people worked. I then typed it out and proceeded to sell them to various magazines, and later to English and American publishers. In two years, I created a legend and doubled J. B. S. Haldane's income."[3] In this retrospective reconstruction of her professional relationship to J. B. S., Charlotte "invents" the scientist as a docile machine, a sort of profitable cyborg, and privileges her ingenuity (in constructing questions for J. B. S. and in marketing his explanations) over the scientist's automatic responses. Clearly, *she* is the creative partner in what is otherwise represented as an essentially mechanistic enterprise.

Yet in spite of this authoritative self-presentation as the controller of the mechanistic scientist, when she assumed the position as editor of *Woman Today,* a socialist woman's magazine published briefly in the late 1930s, Charlotte Haldane adopted a subservient, even adulatory attitude toward scientists and scientific practice, probably because she had then to conform to public opinion in order to keep up the magazine's faltering circulation. Reflecting her awareness that there was an expanding market for scientific knowledge, she built the readership of *Woman Today* by publishing articles that addressed traditional women's magazine topics from a scientific perspective, with an ideological message.

Careful not to alienate her readership, Haldane took a position diametrically opposed to the one she claimed first motivated her to take on the editorial role. Rather than acknowledging her editorial authority in relation to scientists, whether acquaintances or her husband, she played up her own femininity, purveying a traditional construction of the scientist as expert, infallible, and inevitably male. She included scientifically tinged articles reflecting the understanding that scientists, rather than women themselves, were the authoritative sources for knowledge about the female body. And if a scientist who published in *Woman Today* also happened to be a woman, Haldane's editorial treatment of her biography was telling. When she published Dr. Barbara Holmes's essay "The Gland That Controls Your Sex," the biographical note on Dr. Holmes introduced her to readers shorn of her professional title, describing her instead as "Mrs. Barbara Holmes . . . the daughter of a very great scientist, Professor Sir John Gowland Hopkins, a former President of the Royal Society." Even the professional occupation specified for Dr. Holmes was diminished by double domestic modifiers: "Mrs. Holmes works in the Biochemical Laboratory at Cambridge, of which her father is the director" (Haldane, "Calling" 3). Haldane's biographical note affirms patriarchal constructions of the scientist as male, representing women scientists as (merely) members of scientific families—daughters or wives of scientific men. Haldane also cloaked her own narrative of scientific practice within the more acceptable narrative of heterosexual romance, when she published "They Were Two Hours from Death but I Was Not Afraid," her account of a dangerous physiology experiment carried out by her husband in the context of a wartime tragedy—the suffocation of submariners on the *Thetis*. The essay emphasizes (almost to the point of caricature) Charlotte Haldane's domestic, possessive, and hero-worshiping relation to her scientist husband, J. B. S. Haldane:

> The scientific tradition is one of the noblest conventions of mankind. Scientists do not take foolhardy risks. . . . Whenever my husband has made an experiment on himself, for example, he has always told me in advance just what he was trying to found out [sic] and what he thought would happen. *He has never been wrong*. So, as a well-trained scientist's wife, I am not brave, but confident in my husband's ability and knowledge.

This nearly parodic construction of infallible scientist husband and "well-trained scientist's wife"—clearly a capitulation to traditional gender roles—reflects Haldane's strategy of using normal scientific discourse, along with conventional women's magazine topics, to advance radical causes (whether in terms of politics or gender politics). A similar use of traditional gender roles and a traditional construction of science to press nontraditional political positions occurred when *Woman Today* published Leonora Gregory's "Lipsticks are Politics." That essay argues for using "the aids which science has provided to even up the bad deal we may have had from nature and from our environment," in order to deduce a moral that is not biological, but political. "Whereas make-up is frowned on in Germany, lipsticks are in great demand in the Soviet Union," Gregory points out, and she deduces from this contrast "two attitudes to life that go right to the bone. The one: anti-progressive, anti-scientific, which results in restricting the mass of people to a minimum share of what the world has to offer. The other: for utilizing everything that nature and science can provide in order to improve the lot of all" (16–17).

The centrist, meliorist, and patriarchal model for scientific practice that Charlotte Haldane's popular science writing adopted to advance the position of women as wives and mothers contrasts dramatically with the way that Naomi Mitchison used science in her volume *An Outline for Boys and Girls and Their Parents* (1932). Intended as "a group of essays about the state of the sciences and humanities written for children by left-leaning thinkers," *An Outline* caused a controversy upon publication because it included Charles Skepper's critique of the nuclear family as merely "one way of keeping together" (Benton 82–83; Skepper 461–92). Rather than shoring up the family as a way of improving women's position, as Charlotte Haldane's work attempted to do, this essay argued that emphasis on the importance of the family was itself a "condition unfavorable to women" (Skepper 480). "Where the family is important, there is nobody else to look after children, and it becomes all the more necessary that women should be deprived of rights in order to make them specialise in this task. The inferior position of women is therefore . . . closely connected with family life" (480). Skepper's essay in *An Outline* catalyzed a sharp debate in the press: an open letter appeared, attacking the volume for promoting the "break-up of the traditional family," signed by a group of prominent churchmen, headmasters, and other public figures. Left-wing thinkers and writers flocked to Mitchison's defense, among them Dora Russell, Rebecca West, C. E. M. Joad, George Bernard Shaw, Harold Laski, and the publisher himself, Victor Gollancz (Benton 83). The outcome of the controversy ironically recalls Charlotte Haldane's editorial strategy for *Woman Today:* "If we want to run an intelligent paper for intelligent women . . . we must, just like the trusts, go for circulation" (Haldane, "Special Announcement" 9). Sadly, as Benton reports, the "attack on *An Outline* killed it in the market-place. To Naomi Mitchison's disappointment it was never published in the United States," and even in Britain the audience for the volume was so small that Mitchison was unable to pay her contributors (83).

Mitchison's editorial influence was clearly felt in the feminist critique of the family that created so much trouble for *An Outline* owing to the increasing resistance in the 1930s to feminist discourse (Benton 83). But another aspect of Mitchison's work on *An Outline* also merits some examination: her editorial construction of the place of science and the scientist. Unlike Charlotte Haldane's affirmation of "the scientific tradition [as] one of the noblest conventions of mankind" and her self-presentation as the "well-trained scientist's wife," ("They Were Two Hours from Death" 3), Mitchison's editorial preface and her editorial introductions to each author's segment of *An Outline* deliberately relativize the scientific project and decenter the scientist, making both marginal to the central figure in the book: the curious child.

As Mitchison explains to her readers in the "Editorial Preface" to *an Outline:* "this book is planned on a definite scheme. It is all working outward, from Me or You (the one thing of whose existence one is fairly certain) to the Universe. From Now (the present) to all time, past and future" (5). This commitment to making the child central and communicating science as one knowledge practice among many reflects Mitchison's earlier involvement as a contributing editor to the shortlived *Realist: A Journal of Scientific Humanism,* a 1929 publication dedicated to breaking down the disciplinary boundaries separating the sciences and the humanities. Mitchison allows for differences among her readers that will result in different valuations of the different ways of knowing as well: "One kind of knowledge will be exciting to one sort of person, but not to another. If you are bored by any chapter, go on to the next," Mitchison advises (*Outline* 4). In both her choice of illustrations and her text, Mitchison portrays science not as the dominant system of knowledge, but as merely one among many. A series of schematic figures represent the child's relation to the different discursive communities and knowledge practices that make sense of the world for the child (see figures 10.3–10.5). These figures are not hierarchically arranged but rather offer alternative ways to think about the world. Mitchison reinforces that visual point in her "Editor's Preface," where she explains:

> One oughtn't really to separate the history and science parts nearly so definitely, but it made the scheme of the book simpler. . . . Here again, everything fits together, though there are gaps in historical knowledge as there are gaps in scientific knowledge, and it is as exciting to be a good historian as it is to be a good scientist. (*Outline* 9)

Mitchison's casual acknowledgment that disciplinary boundaries are merely schematic conveniences rather than intrinsic truth elements is paralleled, in the biographical paragraphs on her contributors, by the

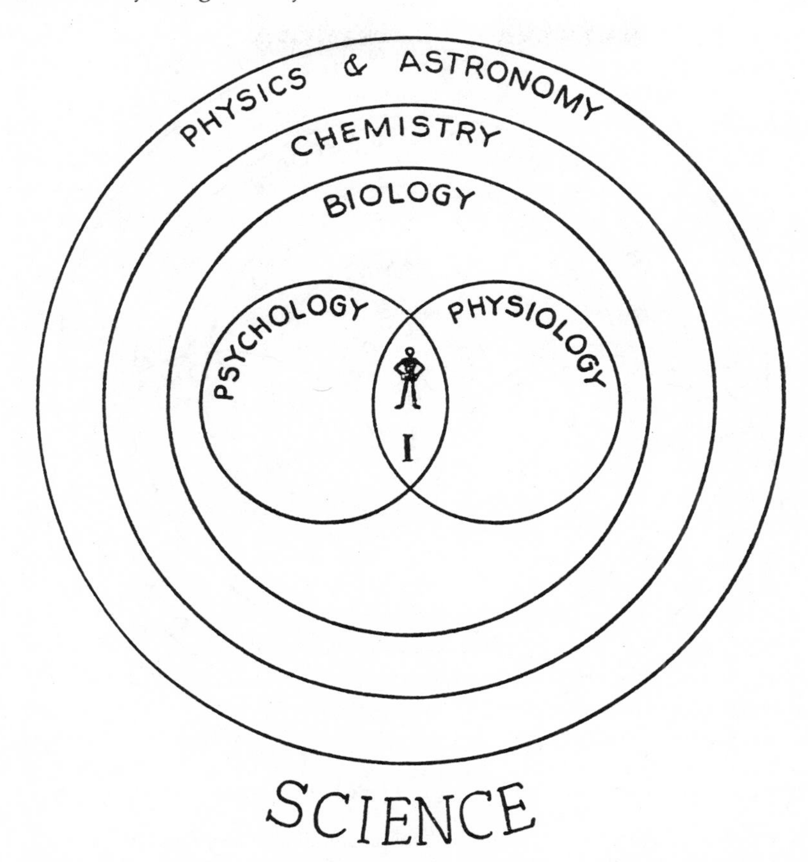

Figure 10.3. Here am "I" in the universe. Illustration from Mitchison's Editor's Preface. Source: *An Outline for Boys and Girls and Their Parents* (London: Victor Gollancz, 1932), 7.

inclusion of personal as well as professional data. Here again, she subverts professional as well as epistemological hierarchies, explaining that "some of my authors are eminent and some are not eminent yet. I have written biographies of most of them, but some of them would not let me do that, and others made me cut out what I thought were the best and funniest bits . . ." (13). Those biographical introductions insist on placing scientific involvement in social context: thus she details the obstacles surmounted by Winifred Cullis, professor of physiology, because "when she was a girl it was not so easy to get a good education and become a doctor"; she mocks Eric Strauss for wearing "an eye-glass, so as to impress patients (or perhaps it really helps him to see them too)"; and she concludes her biography

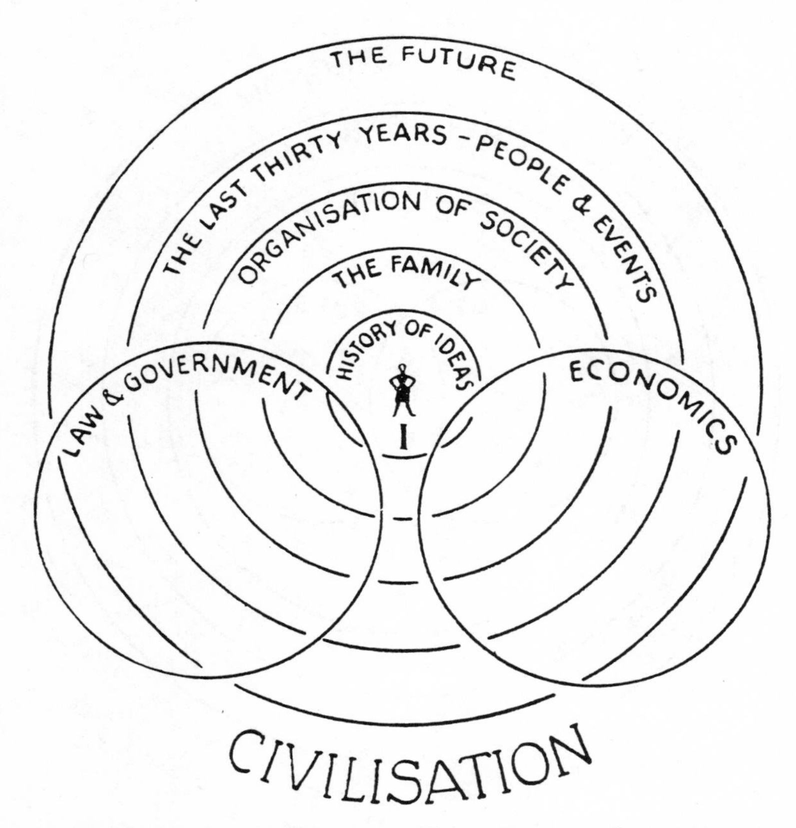

Figure 10.4. Here am "I" in the past and future. Illustration from Mitchison's Editor's Preface. Source: *An Outline for Boys and Girls and Their Parents* (London: Victor Gollancz, 1932), 10.

of N. W. Pirie ("a bio-chemist [who] works in the bio-chemical laboratory at Cambridge, doing experiments") by mentioning, "Just before this book was finished his eldest son, John, was born" (73, 139, 209). Unlike Charlotte Haldane's *Motherhood and Its Enemies,* which began with the ringing assertion of disciplinary—"Any contribution to the discussion of sex problems must, I hold, keep firmly within certain specified limits" (3)—Mitchison's *An Outline for Boys and Girls and Their Parents* revels in the productive transgression of boundaries, whether of disciplines, of genders, or between the public and private realms.

What significance is there in the different ways Charlotte Haldane and Naomi Mitchison practiced feminism in their popular science writings?

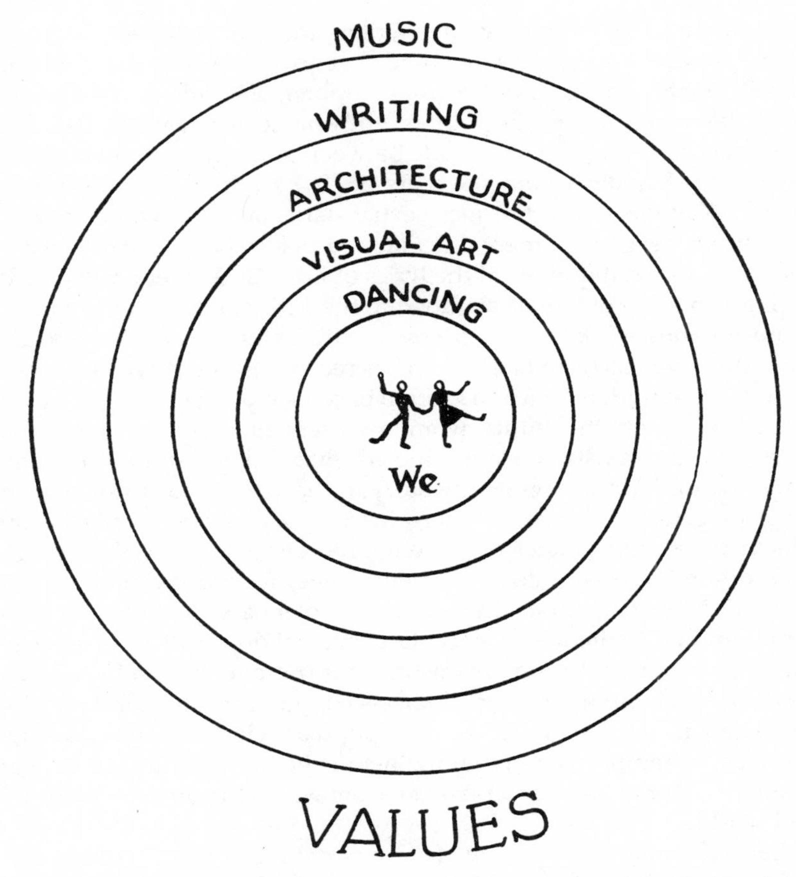

Figure 10.5. Here are "we" in the world of values. Illustration from Mitchison's Editor's Preface. Source: *An Outline for Boys and Girls and Their Parents* (London: Victor Gollancz, 1932), 12.

As with any bit of representational history, the meanings we find are to some degree shaped by our own contemporary concerns: what issues are currently contested and what ideological work *we* are doing in our own writings. I will just sketch out some of the significance I find in the contrast between the science writing of these two modern women, in the hope that readers will add meanings of their own.

Those of us involved in rethinking modernism in terms of a broader notion of its themes and practitioners will find here evidence of a more explicit concern with issues of scientific culture and scientific practice than is usually attributed to modernist writers. Moreover, we will notice that these two modern women writers address the feminist implications of science in a surprising range of genres: encyclopedia, romance fiction, political tract, birth control pamphlet, self-help essay. This generic diversity confirms the point recently made by Suzanne Clark: that no simple correlation can be made between genre choice and ideology in the case of modern women writers (38–39). Far from being the refuge of writers who support the sociosexual status quo, for women writers realistic and even sentimental fiction and poetry have functioned as a site of cultural critique, from the lyrics of Edna St. Vincent Millay with their affirmation of female selfhood to the fiction of Kay Boyle with its representation of politics interwoven with heterosexual love and desire. Moreover, as Ruth Hoberman has recently shown, even historical fiction, often understood as a bastion of conservative masculinism, appealed to twentieth-century women writers as a genre that enabled them simultaneously to claim, and to revise, their cultural heritage. Thus, Laura Riding rewrote the story of the Trojan War to stage a critique of gender-role stereotyping in *A Trojan Ending,* while Naomi Mitchison turned to tales of ancient Greece in Frazer's *Golden Bough,* and to late Victorian tales of ancient Rome, for stories that both expressed the British attraction to imperial power and provided material for a feminist challenge to masculine cultural dominance. In a sense, there is a parallel between Mitchison's incorporation and subversion of classical history in her historical novels and her evocation and revision of science in her science fiction and popular science writing. In each instance, a traditionally masculine field is being entered by a woman writer in order to use it as a platform from which to reimagine both self and world.

Just as a remarkable variety of early-twentieth-century women writers were drawn to historical fiction, so too scholars of early-twentieth-century feminism will find a surprisingly wide range of opinions about scientific practice and the figure of the scientist. Haldane's and Mitchison's writings articulate positions ranging from a meliorist affirmation of normal science to a Latourean challenge to the purified boundaries of science itself. Indeed, in her biographical portraits in *An Outline,* Mitchison gives us the scientist her/himself as a hybrid object—both reproductive body and productive mind, both private parent and public citizen.

Scholars interested in investigating how scientific discourses of the body express resistance as well as power will notice the different ways Haldane and Mitchison used scientific discourse to make arguments on behalf of feminist causes. While both women understood themselves to be working to improve the lot of women, Mitchison did so by challenging the dominance of a male-constructed model of scientific practice, while Charlotte Haldane's call for a scientifically enforced vocational mothering embraced the authority of normal science. Haldane's strategy stands as a poignant example of the "misplaced resistance" that, as political scientist Kathy Ferguson puts it, "appeals to established doctrines for more equitable treatment and thus participates in the perpetuation of those very doctrines" (218). Those of us working in science studies will find in the contrasting writings of Haldane and Mitchison a sample of the many strategies women used to become involved in science: from journalism and essay writing through journal and encyclopedia editing. Finally, those committed to a notion of feminism that is not unitary but multivocal and complex will find in the contrasting science writings of Charlotte Haldane and Naomi Mitchison in the early twentieth century confirmation of the assertion that conflicts—not merely across the gender divide but *in feminism and between women*—are part of the history of feminist response to the problems and promise of contemporary science.

Charlotte Haldane (1894–1969): Major Works:

Motherhood and Its Enemies (1927)
"Special Announcement," *Woman Today* (March 1939)
"Calling All Women," *Woman Today* (July 1939)
"They Were Two Hours from Death but I Was Not Afraid," *Woman Today* (1939)
Truth Will Out (1949)
"My Husband the Professor," BBC Third Programme (11 September 1965)

Naomi Mitchison (b. 1897): Major Works:

Comments on Birth Control (1930)
An Outline for Boys and Girls and Their Parents (1932)
Solution Three (1976)
"All Change Here," *As It Was* (1988)

Notes

1. For analyses of such conflicts, see Hirsch and Keller, especially Mary Childers and bell hooks, "A Conversation about Race and Class": (60–81); and Mohanty.

2. I use the term "protopostmodern" to indicate Mitchison's nonessentialist awareness of difference within feminism, and her consistent stress on the power of location and position in setting a feminist agenda. Similarly, in her science writing she might be described as a "protopostmodern" critic of science

because she understands scientific knowledge as culturally constructed. For a longer discussion of these issues, see my *Babies in Bottles.*

3. Charlotte Haldane, "My Husband the Professor." See *Truth Will Out* (21) for a more decorous version.

Works Cited

Benjamin, Marina. "Elbow Room: Women Writers on Science, 1740–1840, *Science and Sensibility, Gender and Scientific Inquiry, 1780–1945.* Oxford: Blackwell, 1991. 27–59.

Benton, Jill. *Naomi Mitchison: A Biography.* London: Pandora, 1990.

Childers, Mary, and bell hooks. "A Conversation about Race and Class." *Conflicts in Feminism,* ed. Marianne Hirsch and Evelyn Fox Keller. New York: Routledge, 1994. 60–81.

Clark, Suzanne. *Sentimental Modernism.* Bloomington: Indiana UP, 1991.

Eden, Trudy. Rev. of Londa Schiebinger, *Nature's Body: Gender and the Making of Modern Science. Configurations* 2 (Fall 1994): 581–83.

Ferguson, Kathy E. 'Male-Ordered Politics: Feminism and Political Science." in *Idioms of Inquiry: Critique and Renewal in Political Science,* ed. Terence Ball. Albany: SUNY P, 1987. 209–29.

Gregory, Leonora. "Lipsticks Are Politics." *Woman Today,* January 1939, 16–17.

Haldane, Charlotte. "Calling All Women," *Woman Today,* July 1939, 3.

Haldane, Charlotte. *Motherhood and Its Enemies.* London: Chatto & Windus, 1927.

Haldane, Charlotte. "My Husband the Professor." BBC Third Programme, 11 September 1965. National Sound Archive (London).

Haldane, Charlotte. "Special Announcement." *Woman Today,* March 1939, 9.

Haldane, Charlotte. "They Were Two Hours from Death but I Was Not Afraid." *Woman Today,* August 1939, 2–3.

Haldane, Charlotte. *Truth Will Out.* London: Weidenfeld & Nicolson, 1949.

Haldane, J. B. S. *Daedalus or Science and the Future.* London: Kegan Paul, Trench, Trubner, 1923.

Haldane, J. B. S. "How to Write a Popular Scientific Article." *A Banned Broadcast and Other Essays.* London: Chatto & Windus, 1946. 3–8.

Hirsch, Marianne, and Evelyn Fox Keller, eds. *Conflicts in Feminism.* New York: Routledge, 1992.

Hoberman, Ruth. *Resisting History: Classical Antiquity in Twentieth-Century British Women's Historical Fiction.* Forthcoming, SUNY P.

Holmes, Barbara. "The Gland That Controls Your Sex." *Woman Today,* July 1939, 8–9.

Latour, Bruno. *We Have Never Been Modern.* Trans. Catherine Porter. Cambridge: Harvard UP, 1993.

Lewis, Jane. *Women in England, 1870–1950.* Bloomington: Indiana UP, 1984.

Mitchison, Naomi. "All Change Here." *As It Was.* Glasgow: Richard Drew, 1988.

Mitchison, Naomi. *Comments on Birth Control*. London: Faber & Faber, 1930.
Mitchison, Naomi. *An Outline for Boys and Girls and Their Parents*. London: Victor Gollancz, 1932.
Mitchison, Naomi. *Solution Three*. London: Dennis Dobson, 1976, rpt. New York: Feminist P, 1995.
Mohanty, Chandra Talpade. "Feminist Encounters: Locating the Politics of Experience." In *Destabilizing Theory: Contemporary Feminist Debates*, ed. Michele Barrett and Anne Phillips. Stanford: Stanford UP 1992. 74–92.
The Realist: A Journal of Scientific Humanism. 1, no. 1 (April–June 1929). London: Macmillan, 1929.
Schiebinger, Londa. *The Mind Has No Sex? Women in the Origins of Modern Science*. Cambridge: Harvard UP, 1989.
Skepper, Charles. "The Family, or One Way of Keeping Together." In *An Outline for Boys and Girls and Their Parents*, ed. Naomi Mitchison. London: Victor Gollancz, 1932. 461–92.
Squier, Susan. *Twentieth-Century Visions of Reproductive Technology*. New Brunswick: Rutgers UP, 1994.

11

Rachel Carson and Her Legacy

Rebecca Raglon

Few writers have challenged the potent North American ideology of progress as thoroughly and forthrightly as Rachel Carson. Even in such distinguished company as Henry David Thoreau, John Muir, and Aldo Leopold, Carson is notable, for she was more successful than any other writer in alerting the public to the assumptions that were responsible for creating dangerous, irreversible changes in the environment. Thoreau, Muir, and Leopold wrote with the hope that they could persuade their readers to adopt a different ethical stance toward the natural world. Carson wrote with an even greater sense of urgency, showing that arrogant attempts to control the natural world had failed. The fact that pesticides sprayed to control insect populations eventually left their residues in human cells provided horrifying, irrefutable proof that humans and nature are inseparable. Moreover, Carson's 1962 book *Silent Spring* marks a turning point as the supporting role women had played in conservation issues was transformed into a more proactive role. Carson's warning about the dangers of unregulated pesticide and herbicide use was not to be the last; since *Silent Spring*, a long line of women writers, artists, and activists have eloquently highlighted environmental problems ranging from pollution to nuclear disarmament.

There is a bitter irony in remembering Carson not for her poetic

Figure 11.1. Rachel Carson with binoculars. Photo by Erich Hartmann, courtesy of Magnum Photos.

books about the sea but for her anger and her record of nature's abuse. Nature writing is variously defined, but most literary historians view it as a mingling of romantic rapture with an unmediated or scientific viewpoint. Essentially, such writing develops a nonutilitarian viewpoint that opens up the natural world for a sense of enjoyment. Nature writers also tend to see the world primarily as a home for other subjects rather than merely as a storehouse filled with resources for human consumption. As Thomas Lyon points out, "The principal cultural heresy expressed in American nature writing is the refocusing of vision outward from the self, individually, and from the corporate self, our species. A radical proposal follows on the widened vision: the environment, nature, is the ground of a positive and sufficient human joy" (Lyon 19).

At its deepest level, nature writing is joyful and comedic, essentially about life and the continuation of life. Carson was an enthralling writer, as she proves in her three books about the sea, and she was well aware of working within a well-defined tradition. Nevertheless, her greatest legacy comes from another kind of nature book. *Silent Spring* marks the origin of a new kind of nature writing: a dark new genre that deals with the horrific consequences of human actions upon the earth. Carson, like other nature writers, was blessed with the ability to give literary representation to scientific fact. In researching *The Sea around Us,* she estimated that she might have consulted as many as a thousand technical manuals, but felt that her relation to technical scientific writing had been that of "one who understands the language but does not use it" (Brooks 3). Carson vigorously opposed the notion that science was the "prerogative of only a small number of human beings, isolated and priest like in their laboratories." Nor was she willing to accept the idea that words might have an alchemy of their own: in Carson's mind, science and poetry both sprang from the "materials of life." In her 1952 acceptance speech for the National Book Award, she said:

> The winds, the sea, and the moving tides are what they are. If there is wonder and beauty and majesty in them, science will discover these qualities. If they are not there, science cannot create them. If there is poetry in my book about the sea, it is not because I deliberately put it there, but because no one could write truthfully about the sea and leave out the poetry. (Brooks 128)

Carson's career was a complex synthesis of literary achievement, professional scientific training, and government service. It is a career that has been variously interpreted: her first biographer, Paul Brooks, while an admirer of her work, viewed her unmarried status with a certain pity (for most of her adult life, Carson lived with her mother). More contemporary reassessments, however, have tended to look at her life as a full, rich one, and have shown that her career flourished within a culture of women mentors, beginning with her mother and going on to women who befriended her in college, and later in the literary world (Norwood; Hynes; Freeman). Although she frequently found herself in situations that made her a type of "pioneer" for her sex (for example, she was the first woman to go aboard the government research ship *Albatross III*), there is no indication that she was uncomfortable with any aspect of scientific culture until she began work on *Silent Spring*. In fact, the critique Carson was eventually to make of certain aspects of scientific culture could not have been launched unless she had been a successful and knowledgeable participant in that culture (Graham; Hynes).

Much of the work Carson did for the government was educational, ranging from compiling recipes to writing introductory guides for a number of wildlife refuges throughout the United States. Some critics have speculated that this educational emphasis might be responsible for the pedantic tone that creeps into Carson's work from time to time (Gartner 114). It would be unwise for anyone to underestimate or devalue her time spent in government service. Linda Lear points out that in fact Carson was part of a "unique research culture within the government" (29). Carson and her colleagues worked within the one government agency that had a record of concern about the widespread use of pesticides and its effects on wildlife. Thus Carson was very early on in her career privy to information that rarely reached the public. In addition, working within a government bureaucracy provided Carson with indispensable background when she came to sketch out the alliance between government, scientific, and corporate interests that was responsible for the massive and indiscriminate use of pesticides after World War II. In 1964, when testifying before a U.S. Senate committee considering a bill aimed at resolving conflicts between agricultural and wildlife interests, she pointed out that she was not an outsider, but spoke as one who had sixteen years of experience as a government biologist: "I therefore am aware of the problems, the frustrations, the inevitable conflicts that arise when two or more agencies attempt to carry out their sometimes conflicting mandates," she said (Brooks 310).

Carson was born on 27 May 1907 in Springdale, Pennsylvania. Her childhood desire was to become a writer, and her first story, "A Battle in the Clouds," was published in *St. Nicholas Magazine* when she was ten. Later she attended Pennsylvania College for Women (now Chatham College) in Pittsburgh. There she began a major in English, but with the encouragement of Mary Scott Skinker switched fields and graduated with a B.A. in science in 1929. The following summer she was awarded a fellowship at Woods Hole Marine Biological Laboratory. In the fall she attended the Johns Hopkins University on a scholarship for graduate study in zoology, receiving her M.A. in 1932. When her father died in 1935, Carson was forced to end further studies at Johns Hopkins and obtain employment to support her mother and two nieces. She wrote the federal civil service examination, finished first, and became one of two women employed at the Bureau of Fisheries (which merged with the Biological Survey in 1940 to form the Fish and Wildlife Service). Carson worked in the bureau's Office of Information and was responsible for writing and editing numerous government publications. During this period she also returned to her literary interests and began to produce articles for newspapers and magazines. Her first book, *Under the Sea Wind,* was published in 1941.

Some nature writing organizes itself around descriptions of what appear to be unattached, isolated phenomena. Carson, however, was not interested in this type of guidebook presentation. What fascinated her most in nature were intertwined processes, long sweeps of time, the way lives of various creatures developed and intersected, and the continuity of life that persisted through all change. In *Under the Sea Wind,* she respectfully paid tribute to Henry Williamson, her favorite author, by focusing on "characters" in the natural world in much the way Williamson did in *Tanka the Otter.* Carson wrote as her subject demanded: but she was also concerned enough with holding her reader's interest to take the risk of naming the creatures she wrote about. The nonspecialist might not be concerned with the processes that eliminate thousands of mackerel eggs, but readers would be drawn to the fate of the one egg that became "Scomber."

A critic commenting on Carson's work once suggested that like many urban dwellers, Carson had little experience with "uncontrolled nature" and consequently did not know how "unpleasantly hostile it generally is" (*Time,* 24 April 1964). To suggest that Carson was unaware of natural processes, however, is a distortion of her work. She was clearly aware of the predator-prey relationships occurring within the wild, and understood them thoroughly enough not to recoil from them. On the other hand, characterizing nature as unpleasant and hostile has been a time-honored technique used to justify the many types of atrocities humans have directed against the natural world. Poets, novelists, and social philosophers have all allowed themselves to become fixated from time to time by the horrors of the natural world. Nor is this fixation strictly a nineteenth-century phenomenon. Annie Dillard, the Pulitzer Prize–winning author of *Pilgrim at Tinker Creek* (1974), finds the fecundity and casual wastefulness of the world appalling: she is horrified by the ants that "take to the sky in swarms" or rivers that run "red and lumpy with salmon" or by the "milky clouds" of barnacle larvae that hatch in the sea (169). Such phenomena inspired Carson, rather than repelled her: the eggs and milt of the mackerel, for example, are described as a "sprawling river of life, the sea's counterpart of the river of stars that flows through the sky as the Milky Way" *Under the Sea Wind* (112). Dillard is disturbed by the lack of choice in nature, by the fact that "fish gotta swim and bird gotta fly; insects, it seems, gotta do one horrible thing after another" (65). Carson, on the other hand, is deeply touched by the "fixity" that Dillard finds so horrifying. In *The Sea around Us,* Carson describes the life of a small sea worm, Convoluta, which has a symbiotic relationship with a form of algae. In order that photosynthesis may occur in the algae, these sea worms rise from the sand of the intertidal zone at each ebb tide. The whole life of this tiny worm is attuned to the movement of the tides:

> What I find most unforgettable about Convoluta is this: sometimes it happens that a marine biologist, wishing to study some related problem, will transfer a whole colony of the worms into the laboratory, there to establish them in an aquarium, where there are no tides. But twice each day Convoluta rises out of the sand on the bottom of the aquarium, into the light of the sun. And twice each day it sinks again into the sand. Without a brain, or what we so would call a memory, or even any very clear perception, Convoluta continues to live out its life in this alien place, remembering in every fiber of its small green body, the tidal rhythm of the distant sea. (166)

It was in small creatures such as the Convoluta that Carson was most apt to find the symbols she needed to express the dual strands of fragility and strength making up the complexity of the world. Carson's own choice for a symbol of life was a small crab juxtaposed with the immensity of the sea: "From my own store of memories, I think of the sight of a small crab alone on a dark beach at night, a small and fragile being waiting at the edge of the roaring surf, yet so perfectly at home in its world. To me it seemed a symbol of life, and of the way life has adjusted to the forces of its physical environment" (Brooks 315).

There are humans in *Under the Sea Wind,* but their role is minimal. They are fishermen, who wait for returning schools of mullet, or for mackerel. One fisherman, still new to his work, sometimes thinks about the fish he is pursuing: "What had the eyes of the mackerel seen? Things he'd never see; places he'd never go. He seldom put it into words, but it seemed to him incongruous that a creature that had made a go of life in the sea, that had run the gauntlet of all the relentless enemies that he knew roved through the dimness his eyes could not penetrate, should at last come to death on the deck of a mackerel seiner" (200).

Throughout *Under the Sea Wind* there is an emphasis on the sweep of life: in the way an eel migrates, a mackerel grows, a bird mates. Each life is filled with incident, touching on innumerable other lives. Together these myriad lives form a harmony of movement and time that contradicts the way humans parcel out the significant incidents of their lives. The opening pages deal with the many lives circulating around a small island, eating and being eaten in turn. Yet the overall effect is one not of brutality, but of harmony. Carson frequently alludes to the most delicate type of music reverberating through the natural world: "It would have taken the sharpest of ears to catch the sound of a hermit crab dragging his shell house along the beach just above the water line: the elfin shuffle of his feet on the sand, the sharp grit as he dragged his own shell across another, or to have discerned the spattering tinkle of the tiny droplets that fell when a shrimp, being pursued by a school of fish, leaped clear of the water" (14). Even in soundless movement interrelationships are manifest. When a school of mackerel several miles long passes through the water "like molten metal," schools of "launce and anchovies must feel the vibrations . . . and hurry in apprehension through the green distances of the sea; and it may be that the stir of their passage is felt on the shoals below—by the prawns and crabs that pick their way among the corals, by the starfish creeping over the rocks, by the sly hermit crabs, and by the pale flowers of the sea anemone" (111). The rhythms of this larger time are an underlying concern in all her books, one she returns to again and again. It is her way of reminding her readers that humans are like the fisherman aboard the seiner, peering down into the depths, but able to see and understand only a fraction of what moves below the surface.

In her second book, *The Sea around Us,* Carson focuses intensely on the idea of eternal rhythms—the continuity that persists through change. The book is deliberately biblical in tone, beginning with an opening quote from Genesis. But here there is no sky god overseeing creation. Carson reworks the well-known metaphor "mother earth," transforming it into the more appropriate "mother sea." All life forms, from small microorganisms to complex mammals, are literally the children of the sea. Again Carson focuses on the links that exist between all forms of life. For example, when animals finally did move to land, "they carried with them a part of the sea in their bodies, a heritage which they passed on to their children and which even today links each land animal with its origin in the ancient sea. Fish, amphibian, and reptile, warm-blooded bird and mammal—each of us carries in our veins a salty stream in which the elements sodium, potassium, and calcium are combined in almost the same proportions as in sea water" (14). Over centuries life has migrated to and from the sea. According to Carson, even humans returned to the sea—at least imaginatively, using their ingenuity to explore and investigate the remotest parts of the ocean. And yet humans return to "mother sea" only on her terms. Carson takes evident satisfaction in the knowledge that man "cannot control or change the ocean, as in his brief tenancy of earth, he has subdued and plundered the continents" (15).

It is difficult to imagine any other writer being able to convey such enthusiasm over the process of sedimentation, describing it as the "epic poem of the earth." Dust, sand, gravel, pebbles, fragments of iron and nickel, and billions of tiny shells and skeletons all become flakes in this great snowfall. Carson confesses to having her imagination set afire by thinking of this "unremitting, downward drift of materials from above, flake upon flake, layer upon layer—a drift that has continued for millions of years, that will go on as long as there are seas and continents" (74). All through *The Sea around Us,* Carson is drawn to cycles so vast they cannot be perceived by human senses, but must be imagined. From a "Stygian world of heated rock and swirling clouds and gloom" (6) to the stories of lost continents, Carson speaks to the imaginations of her readers and to their perennial fascination with the sea. *The Sea around Us* was a popular book, selling well enough and earning enough to allow Carson to quit her government job and devote herself full time to writing.

In order to understand Carson "the environmentalist," it is necessary to understand this first part of her career, and the profound respect she held for all life. Nature writers of an earlier era, both men and women, frequently derived moral precepts from nature, and could easily idealize the family life of song birds while condemning the predations of the hawk. Carson, writing at a time when such moralizing had dropped out of favor, saw in each living thing its own peculiar beauty. She had respect even for those awkward imports into North America, starlings and rats. She saw no contradiction in loving both cats and birds, and she expressed pain and outrage when she witnessed the men aboard a government research ship shooting at the sharks that she found so beautiful. She understood that even the insects that humans found so discomforting were actually an essential part of life.

After finishing her books on the sea, Carson planned to write a book dealing with the origin of life but was directed to the issue of pesticides when she realized that "everything which meant most to me as a naturalist was being threatened" (Brooks 233). There was pain in learning that what she had held most dear in her scientific and artistic life was threatened. The "eternal truths" that she had written about—the harmony of the natural world, the long rhythms of a life cycle, the interconnections existing between sea and land, wind and tide, plant and animal—were all under attack. *Silent Spring* originated as her way of coming to terms with the man-made changes engulfing the world. In a 1958 letter Carson wrote:

> I have been mentally blocked for a long time, first because I didn't know just what it was I wanted to say about Life, and also for reasons more difficult to explain. Of course everyone knows by this time that the whole world of science has been revolutionized by events of the past decade or so. I suppose my thinking began to be affected soon after atomic science was firmly established. Some of the thoughts that came were so unattractive to me that I rejected them completely, for the old ideas die hard, especially when they are emotionally as well as intellectually dear to one. It is pleasant to believe, for example, that much of Nature was for ever beyond the tampering reach of man: he might level the forests and dam the streams, but the clouds and the rain and the wind were God's . . . . These beliefs have almost been part of me for as long as I have thought about such things. To have them even vaguely threatened was so shocking that, as I have said, I shut my mind—refused to acknowledge what I couldn't help seeing. But that does no good, and I have now opened my eyes and my mind. I may not like what I see, but it does no good to ignore it, and it's worse than useless to go repeating the old "eternal verities" that are no more eternal than the hills of the poets. So it seems time someone wrote of life in light of the truth as it now appears to us. (Freeman 248–49)

As she herself knew, it would have been much more pleasant to go on writing about the beauty of nature, but knowledge of the depredations occurring forced her research into another direction. Friends wrote to her, describing their own distress upon seeing birds dying in agony after spraying with DDT. Nor was there any lack of technical information available in the 1950s about the dangers of pesticides to human and animal health. Carson's accomplishment was in synthesizing information from a wide variety of sources, and presenting it as an issue that touched on every aspect of life, including human life. In doing this, she clearly transcended the limited confines of the average scientific inquiry.

As Frank Graham points out, because of Carson's stature as a best-selling author and public figure, the pesticide controversy could no longer be confined to technical journals but entered public debate. But Carson did more than alert the public to a critical problem. According to Graham, "she uncovered and pointed out publicly for the first time, even to many scientists, the facts which link modern contaminants to all parts of the environment. There are no separate environmental problems . . . she synthesized the issue—for the scientists, the public, and the government" (x).

Carson began her career as a traditional nature writer, but her greater significance rests with her critique of the scientific, governmental, and industrial actions collectively seeking to "manage" the natural world. Thirty years after the publications of *Silent Spring*, she remains one of the most eloquent critics of this enterprise:

> The "control of nature" is a phrase conceived in arrogance, born of the Neanderthal age of biology and philosophy, when it was supposed that nature exists for the convenience of man. The concepts and practices of applied entomology for the most part date from that Stone Age of science. It is our alarming misfortune that so primitive a science has armed itself with the most modern and terrible weapons, and that in turning them against the insects it has also turned them against the earth. (*Silent Spring* 297)

Carson relied on technical journals, anecdotal evidence, and correspondence with a variety of experts around the world in creating her synthesis. But philosophically, what underlies her work is what Paul Brooks identified as a "central intellectual conflict . . . the conflict between those scientists who are willing to extrapolate possibilities from the known facts, and the 'positivists' who say there can be no damage because damage has not been demonstrated" (241). Carson herself hinted that there might be a gender bias involved when she wrote: "I'm convinced there is a psychological angle in all this: that people, especially professional men, are uncomfortable about coming out against something, especially if they haven't absolute proof that 'something' is wrong, but only a good suspicion. So they will go along with a program about which they privately have acute misgivings" (Brooks 241).

Carson's book also critiqued the myth of pure science. Clearly, scientists could be influenced just like other humans—by grants, employment, prestige. When a chemical company offered such inducements to scientifically trained men, the type of science that resulted was indeed suspect:

> It was reported in 1960 that only 2 per cent of all the economic entomologists in the country were working in the field of biological controls. A substantial number of the remaining 98 per cent were engaged in research on chemical insecticides . . . . Why should this be? The major chemical companies are pouring money into the universities to support research on insecticides. This creates attractive fellowships for graduate students and attractive staff positions. . . . This situation also explains the otherwise mystifying fact that certain outstanding entomologists are among the leading advocates of chemical control. Inquiry into the background of some of these men reveals that their entire research program is supported by the chemical industry. Their professional prestige, sometimes their very jobs depend on the perpetuation of chemical methods. Can we then expect them to bite the hand that literally

feeds them? But knowing their bias, how much credence can we give to their protests that insecticides are harmless? (*Silent Spring* 258–59)

That such concerns continue to inform many scientific controversies is clear from contemporary debates dealing with genetic engineering and reproductive technology. While the metaphors used today are different (bio-entrepreneurs are "assisting" rather than "controlling" nature), and the substances are different (an "ice-minus" bacteria is sprayed over crops rather than DDT), the outlines of the debate have remained disquietingly similar. Once again certain scientific, business, and government interests are promoting the use of technologies that will enhance economic performance on the basis that not only has there been no "proven" danger, but in fact humans will reap the benefits of greater crop yields, more milk production, more human babies. Carson was able show the links between science and industry in *Silent Spring*, and similar alliances continue to drive the type of research done in the fields of reproductive and biotechnologies.

As a synthesizer, Carson realized that problems could not be solved in isolation. In *Silent Spring*, Carson shows how initial successes frequently were followed by unexpected consequences: the insects that were to be controlled by the spray gun became resistant to chemicals. Since the processes of nature are infinitely more long ranging and complex than humans can even imagine, Carson believed that a more sophisticated attitude toward scientific and technological invention was warranted. Yet chemical weed killers give such "a giddy sense of power over nature to those who wield them, [that] the long-range and less obvious effects . . . are easily brushed aside as the baseless imaginings of pessimists" (*Silent Spring* 69).

While relying heavily on technical journals to build her case against the scientific, chemical, and government interests that favored massive spraying programs, she gathered a great deal of the anecdotal evidence in her book from nonexperts—suburbanites and ornithologists upset by the dead and dying birds they found on their lawns and college campuses, campers poisoned by DDT, grandmothers protesting the "scorched earth" policy practiced along roadsides. In utilizing this testimony she picked up a strand of thought she had investigated in an article eventually published as *A Sense of Wonder* (1965). In that small book Carson discusses the ways a parent can help a child develop a love of nature. Carson reassured her readers that investigating the natural world is not an activity that should be turned over to professionals, nor was it necessary to have "all the facts." According to Carson, it was far more important to convey to children feelings for nature, rather than lifeless bits of information.

This respect for the amateur was also an essential element of *Silent Spring*. Reports from a woman returning home from church who saw an "alarming number of dead and dying birds" (90) bracket the almost unthinkable idea that decisions made by a variety of "experts" were adversely effecting the day-to-day lives of Americans. The fact that Carson relied on these nonexpert witnesses when weaving her narrative together is significant: after *Silent Spring* scientific expertise would no longer be a guarantee of public confidence and trust, and environmentalism would be conceived as a social movement with a broad and far more democratic base than that of the conservation era that preceded it (Hays).

Women have long been involved in certain aspects of nature study, but Carson turned a genteel tradition on its ear. While it was once considered a pleasant pastime for women to collect seashells or ferns, or to illustrate bird or plant life (Barber), Carson's legacy has insured that such innocent nature appreciation will now have to occur within a much darker context. Some feminist writers are dismayed with any suggestion that women might have a "special" connection with nature. Carson's type of writing, however, ensures that much of this intellectual distress is moot. Women, men, and children have all discovered in the late twentieth century that their bodies are not only connected to nature, but an inescapable part of it. Thus when nature is perceived as scarred and abused, the implications for humans become frightening. As a result of Carson's work, the descendants of those Victorian shell-and-fern collectors, those lady botanists and bird-watchers, today find themselves scrutinizing a mutant nature that includes themselves. This unfortunately is also the Carson legacy, and it is to be found in a host of contemporary dystopic nature books, films, photographs, and paintings by women.

For example, the work of Cornelia Hesse-Honegger, a Swiss scientific illustrator, is particularly interesting when looked at in the context of Carson's legacy, for Hesse-Honegger, like Carson, has turned her careful glance upon the insect world. Concerned by the lack of knowledge about the effects of low-level radiation, Hesse-Honegger has toured nuclear power plants around the world in order to paint the insects found in their precincts. The resulting paintings are both beautiful and horrifying: beautiful upon first glance, intricate in detail and glowing with rich colors, yet horrifying when viewed more closely and it is possible to see the missing antenna, the strange growths, asymmetrical body parts, and stunted wings. One has an aesthetic reaction similar to that of reading Carson's work: a sense that the familiar world has just shifted and something ominous has taken its place, something that cannot be ignored. Hesse-Honegger, as if to underscore the need to really see, paints her insects larger than life.

After a special trip to Sweden to study the area that received the highest fallout from Chernobyl, Hesse-Honegger published some of her findings and was criticized by the scientific community. Again the chief complaint was that anecdotal evidence such as that found in her paintings does not constitute serious "proof" that low-level radiation is the cause of such mutations. Stung by the criticism, Hesse-Honegger began making more comprehensive collections, beginning with the countryside around Swiss nuclear plants. Fully expecting to find less damage than in areas where accidents occurred, she was struck by the numbers of greatly deformed insects she found. In 1989 she spent her summer holidays collecting 434 insects around the nuclear waste reprocessing plant at Sellafield, England, finding the most damaged insects in the proximity of the Sellafield complex. In 1990 she visited Chernobyl itself, and in an area west of the exclusion zone almost every insect she found was mutated. (In contrast the insects she found around Kiev were normal.) In 1991 she visited the Three Mile Island plant and again found all the "familiar" mutations (Norman 101–3).

Marilynne Robinson is an American novelist who has also turned her attention, in the Carson tradition, to the study of the consequences of human actions upon the natural world. Robinson is the author of the justly acclaimed novel *Housekeeping,* which is filled with lyrical descriptions of the natural world. In *Mother Country* (1989), however, Robinson finds herself being drawn into a study of the harm being done to the natural world, examining in particular the cultural assumptions that make the English nuclear reprocessing plant of Sellafield possible. She suggests in *Mother Country* that the genial portrayal of England as a nation of "nature lovers" is little more than protective coloring to hide one of the most abused landscapes in the industrialized world. As Carson did in *Silent Spring,* Robinson traces the links between scientific and government interests to find an explanation for the environmental degradation. Carson questioned whether a society given over to the indiscriminate killing of wildlife could be called civilized; Robinson asks the same question of Great Britain, a fastidious nation excessively concerned about rabies yet willing to import tons of dangerous chemical and radioactive waste. Lying somewhere within the inquiries of both women is a radical questioning of authority—the authority of professionalism, the authority of governments, and the authority of cultural convictions. Robinson's book is an attempt to question everything, to jar loose the reasons a government would adopt a policy toward its land and its own population of a "shrewd adversary contriving to do harm for profit" (31). Much of Robinson's outrage stems from the fact that Sellafield is no secret, and what she writes about can be found on the front pages of British newspapers: accounts of toxic waste "weeping" into streams, of spills, slicks, beach closures and cleanups, extinctions, and elevated cancer rates. While on one hand plutonium is described as one of the most dangerous substances in the world (129), on the other hand, according to government publications, there is no danger in pumping plutonium wastes into the North Sea. As Robinson comments, "all this deserves its own chapter in any history of modern thought, simply because its consequences are epochal" (128).

In *Refuge,* another book in the Carson tradition, the American biologist Terry Tempest Williams draws together the complex threads of events that made her, at the age of thirty-four, "the matriarch of my family" (3). Williams, who works as a naturalist at a Utah museum, interweaves the events in her family with the inexorable rise of the Great Salt Lake in the 1980s, and the threat this posed to a bird refuge. The rise of the lake was a natural occurrence, but the lack of alternative places for birds and wildlife to retreat to was not "natural." Williams draws parallels here between the inability of the threatened birds and animals to find another "safe" home, and the lack of a "safe" home for Utah's human inhabitants.

While the Atomic Energy Commission described the country north of the Nevada Test Site as virtually uninhabited desert terrain, Williams notes that "my family and the birds at Great Salt Lake were some of the 'virtual uninhabitants' " (287). All through her life she has been haunted and intrigued by a dream about a flash of light in the desert. Only much later does she discover that the flash of light was part of Operation Plumbbob, and that the "dream" occurred on a September night in 1957 when her family happened to be traveling through the desert where testing occurred. "I cannot prove my mother, grandmothers, aunts . . . developed cancer from nuclear fallout in Utah. But I can't prove they didn't" (286). Williams also points out that her cultural heritage is also responsible for allowing atmospheric testing to occur: "In Mormon culture, authority is respected, obedience is revered, and independent thinking is not. I was taught as a young girl not to 'make waves' or 'rock the boat' " (285). Williams's book concludes with an act of civil disobedience by a dozen women who decide to "rock the boat." As they cross over into the Nevada Test Site, Williams feels that a "new contract was being drawn by the women, who understood the fate of the earth as their own" (288).

Williams's and Robinson's voices, like Hesse-Honiger's drawings, graphically dispute the cozy reassurances of the experts and regulators who insist that there are "acceptable levels" of contamination, or that no harm is being done because no harm can be proven. They contest the legitimacy of scientific evidence and scientific knowledge based on standards of proof and philosophical assumptions that defy common sense. Finally they insist on the legitimacy of their own knowledge, experiences, and feelings, and that, ultimately, may be Rachel Carson's most lasting and important legacy.

Rachel Carson (1907–1964): Major Works

Under the Sea Wind (1941)
The Sea Around Us (1951)
The Edge of the Sea (1955)
Silent Spring (1961)
A Sense of Wonder (1965)

Works Cited

Barber, Lynn. *The Heyday of Natural History.* New York: Doubleday, 1980.
Brooks, Paul. *The House of Life.* Boston: Houghton Mifflin, 1972.
Carson, Rachel. *The Edge of the Sea.* Boston: Houghton Mifflin, 1955.
Carson, Rachel. *The Sea around Us.* New York: Oxford UP, 1951.
Carson, Rachel. *A Sense of Wonder.* New York: Harper & Row, 1965.
Carson, Rachel. *Silent Spring.* Boston: Houghton Mifflin, 1961.
Carson, Rachel. *Under the Sea Wind.* New York: Simon & Shuster, 1941.
Dillard, Annie. *Pilgrim at Tinker Creek.* New York: Bantam Books, 1974.
Freeman, Martha, ed. *Always, Rachel: The Letters of Rachel Carson and Dorothy Freeman, 1952–1964.* Boston: Beacon P, 1995.
Gartner, Carol. *Rachel Carson.* New York: Frederick Ungar, 1983.
Graham, Frank, Jr. *Since Silent Spring.* Boston: Houghton Mifflin, 1970.
Hays, Samuel P. *Beauty, Health, and Permanence.* Cambridge: Cambridge UP, 1987.
Hays, Samuel P. *Conservation and the Gospel of Efficiency.* Cambridge: Cambridge UP, 1968.
Hynes, H. Patricia. *The Recurring Silent Spring.* New York: Pergamon P, 1989.
Lear, Linda. "Rachel Carson's Silent Spring." *Environmental History Review* 17, no. 2 (1993): 23–48.
Lyon, Thomas. *This Incomparable Lande.* Boston: Houghton Mifflin, 1989.
Norman, Geraldine. "Vision of a Mutant Age." *Independent on Sunday,* 10 October 1993, 100–103.

Norwood, Vera. 'Heroines of Nature: Four Women Respond to the American Landscape." *Environmental Review* 7, no. 1 (1984): 34–55.
Norwood, Vera. *Made from This Earth.* Chapel Hill: U of North Carolina P, 1993.
Robinson, Marilynne. *Mother Country.* New York: Ballantine Books, 1989.
Williams, Terry Tempest. *Refuge: An Unnatural History of Family and Place.* New York: Pantheon Books, 1991.

PART 7

SELF-FASHIONING

12

The Spectacle of Science and Self
Mary Kingsley

Julie English Early

Mary Kingsley somewhat regularly skirmished with her publisher, Macmillan, during the preparation of *Travels in West Africa* (1897). Macmillan was confused by its narrative voice, and unsure of its serious or comedic intent; he thought readers would even be uncertain of its author's gender. Kingsley protested, "it does not matter to the General Public what I am as long as I tell them the truth" (Macmillan Letters, 18 December 1894). Kingsley's optimism about the General Public, what she knowingly called the G.P., was realized: a little more than a year later, her book "took the world by storm" (Smith and Ward 349). Yet the confidence she expressed was also disingenuous; the generosity of the G.P. would have to be cultivated.

In the year before publishing *Travels in West Africa,* Kingsley consciously produced and managed a public presence on the lecture platform and in the periodical press that established her not only as a naturalist, an ethnographer, and an observer of West African affairs, but as a distinctly unusual one. Long before her book appeared, Mary Kingsley was an event: a vastly entertaining, sometimes puzzling, and often controversial self-performer—who also laid claim to serious science. This incongruous and compelling public presence guaranteed that *TWA* would be reviewed in publications across the broadest spectrum: from *Punch* to the Royal Geographical Society's *Geographical*

Figure 12.1. Mary Henrietta Kingsley (1862–1900). Studio Portrait c. 1897 by A. E. Hull, London. By courtesy of the National Portrait Gallery, London.

Journal, from the *Illustrated London News* to the *Edinburgh Review*. But if Kingsley's unusual public presence guaranteed attention to her work, it also foregrounded her pointed disorientation of conventional expectations in which the image of the proper scientist stood for the value of the science presented. In her person and in her work, Kingsley stood for reconfigurations that could potentially forestall calcification of disciplines mistakenly confident that their "inherent" objectivity and remote, respectable demeanor were signs of substance. Her methodological and textual self-consciousness pointed to theoretical and political concerns that anthropology would only slowly come to address.

In creating her public presence, Kingsley was astute about herself, her material, and the politics of the sciences at the end of the century. Wisely understanding opportunities for the unconventional contributor to the still loosely formed disciplines of the human sciences, she brought her anthropological fieldwork rather than the work of the naturalist directly to the public. Kingsley limited discussion of collecting and classifying natural specimens to direct exchanges with the British Museum, which, however, provided her with approval in an adjacent discipline. *TWA*, for example, includes plates of the "new" fishes named for her, and, as two of the text's five appendices, the museum reports on her finds. When she received them, she wrote Macmillan with details of the value of her collection, adding, "these things ought to shed a sort of glow of respectability over me" (16 February 1896).

Credentialed only by an impressive intellect, extensive independent study, and firsthand experience, Kingsley lacked an institutional imprimatur and recognized the value of that "sort of glow." The museum's approval secured her a place in the respectable world of species and genera, but as a borrowed glow, also gave her a position from which to challenge a tradition of gentlemanly good science. A clearly defined model of achievement in which the clubbish worth of the individual was collapsed with the value of "his" work shaped, for example, the Royal Geographical Society—a capacious institutional umbrella for all travel-related study. At mid-century, when RGS president Sir Roderick Murchison added public lectures by notable nonmembers, some found the measure too liberally enthusiastic: "[his] popular methods in the reception of the lion of the hour . . . were distasteful to some of the great men of science. . . . There was some uneasiness even amongst his friends" (Mill 80).

At the end of the century, even with pressures on professional societies to adjust their perspectives,[1] Kingsley had good reason to use whatever informal certification she could obtain.[2] Her disadvantages in the face of the gentleman's respectable model were nearly all-encompassing. She was a woman; she had not been formally educated; and, despite the Kingsley name, she did not fully embrace the preferred demeanor of class and/or the professions. As she wrote Macmillan, "I am afraid you have taken up with a complicated criminal" (16 February 1896). Excluded from formal channels of learning, Kingsley, like many women, was a gifted autodidact who educated herself from her father's library, a quixotic collection that may have reflected the individuality of its owner but could prove a treacherous guide for his self-tutored daughter. She recounted that as a young woman

> I happened on a gentleman who knew modern chemistry and tried my information on him. He said he had not heard anything so ridiculous for years, and recommended I should be placed in a museum as a compendium for exploded chemical theories, which hurt my feelings very much and I cried bitterly at not being taught things.[3]

Nonetheless, Kingsley had greater experience of her father's library than she had of her father, who existed principally in entertaining letters sent sporadically from his travels around the world. Kingsley's daily influences were not the famous Kingsley relatives, but her mother, a cockney cook whom George Kingsley had married four days before Mary's birth, and the Baileys, her mother's working-class family. At home, Kingsley gained a vocabulary and a cadence to her speech that stood her in good stead with Liverpool's West African traders; but when she moved in other circles, she would choose to control (or not to control) certain tendencies. One newspaper review of a lecture complained of her fashionable gesture of dropping *g*'s, a criticism that amused her "when I am trying so hard to hold on to the 'h's' " (unidentified letter quoted in Frank 24). Her conversational informality was equally an issue. During the preparation of *TWA*, Macmillan's editorial consultant, Henry Guillemard, had consistently attempted to "professionalize" her language and to cut her stories short. She had just as consistently resisted. In the preface, Kingsley parodies the style she has refused, demonstrates her own agility in moving among styles, and takes full responsibility for her choice: "It is I who have declined to ascend to a higher level of lucidity and correctness of diction than I am fitted for" (viii).

In the flux of social and professional redefinitions of the '90s, Kingsley did not attempt to disguise her disadvantages, but instead plainly saw their potential for disturbing existing hierarchies. She wrote to her friend Lady Macdonald:

I am really beginning to think that . . . the person who writes a book and gets his FRGS [Fellow of the Royal Geographical Society] etc, is a peculiar sort of animal only capable of seeing a certain set of things and always seeing them the same way, and you and me are not of this species somehow. What are we to call ourselves? (undated letter quoted in Gwynn 131)

Kingsley, another "species" who neither would see only "a certain set of things," nor always "[see] them the same way," built upon the traditional collapse of the person and the work to redefine its implications and possibilities. Whether the "lion of the hour" or, as she put it, "the sea-monster of the season," Kingsley called attention to herself as much as to her material, even multiplying and exaggerating the marginalities she could represent to elude a too swift classification in any one of them.[4]

Just as she would caution against the arrogance of too readily reading the cultures of West Africa, so did she subvert any easy readability of Mary Kingsley. A rather slight figure in her mid-thirties, she appeared on the lecture platform in somber black silks several decades out of fashion. Introducing herself at one gathering, she suggested, "I expect I remind you of your maiden aunt—long since deceased."[5] The archaic chapter headings to *TWA*, too, speak in the voice of another age: Chapter IV, "Which the general reader may omit as the voyager gives herein no details of Old Calabar or of other things of general interest, but discourses diffusely on the local geography and the story of the man who wasted coal" (73). Parodic, pointedly anachronistic, even dandyish, the device recalls a time when the dandy was the "natural" province of the leisured aristocrat, but is now one of the many modes that the woman and her text may, at will, dress up in. Any presuppositions suggested by her "quaint but modest appearance" (*Advertiser and Exchange Gazette* (Hull), 13 November 1897, quoted in Frank 246) and archaic respectability were confounded by her "unladylike" views: Kingsley, for example, supported the liquor traffic, deplored the activities of missionaries, preferred the company of the "palm-oil ruffians," the West Coast traders, and named as her favorite West African tribe, the cannibal Fan. With grim flippancy, Kingsley portrayed herself as an anachronism—a survivor of West Africa, the White Man's Grave; a woman with experiences in pursuit of scientific study as thrilling as those of Mungo Park and Richard Burton in pursuit of geographic knowledge and conquest; and a woman, seeming oddly out of time and place, with the riveting drama, wit, and timing of a masterful (and generally masculine) raconteur.

Kingsley's hyperbolic and flippant humor served as the readiest characteristic for her critics to fasten onto, offering them the opportunity for what appeared even willful misunderstanding. She wrote Macmillan following a lecture early in her public appearances: "The Scotch seem to have on the whole understood me perfectly, not so some distinguished English friends who are now attacking me for speaking flippantly on cannibalism" (16 February 1896). Kingsley's discursive storytelling was, in fact, tightly controlled, relying on impeccable timing, making its points by indirection, surprise, exaggeration, and often irony—a narrative style that fully controls its persona, its material, and its audience. At the beginning of their correspondence, Kingsley wrote Professor and Mrs. E. B. Tylor: "I very humbly beg to plead that statements I make seemingly light-heartedly have really had put into them weeks, sometimes months, of very hard work. . . . It is my apology and I know many things I am going to publish require an apology" (1 October 1896).[6] Her style on the platform and later in her books—what reviewers frequently called "racy"[7]—brought thousands to her dramatic lantern slide lectures; audiences of 1,800 to 2,000 were not unusual.[8] Audiences of the 1890s with a culturally induced appetite for spectacle were little concerned with what to make of her, but instead found her gratifyingly informative and entertaining.

Unlike the General Public, scientific and professional observers were troubled by their inability to explain a woman who cavalierly ruptured the connection between sober science and the demeanor of the scientist. Even Tylor's obituary memoir, while it praises her work, betrays his faint puzzlement over how to account for it:

> Some of Mary Kingsley's readers may have been led astray by her light chaffy style into calling her superficial. . . . During the few years I had the privilege of her friendship, I came to appreciate her power of getting to the back of the negro mind. In her own peculiar way she will hardly be replaced. (Green 7)[9]

Incongruent not only with the demeanor of the proper scientist, Kingsley's self-presentation, "her own peculiar way," prompted an array of defensive memoirs that sought to construct a "normative" woman behind the screen of her talent for farce, disconcertingly deadpan tales, and uncomfortably pointed wit. Alice Green, for example, stressed domestic skills: "She was a skilled nurse, a good cook, a fine needlewoman, an accomplished housewife" (3). Dennis Kemp's memoir for the *London Quarterly Review* stressed devotion to her younger brother, Charley ("a most beautiful love-story might be told" [143–44]); their relations were, in fact, strained. As she had asked, "What are we to call ourselves?"

Decidedly captivating and decidedly odd, Kingsley was clearly adept at performing herself, but also at performing science. Kingsley's critics perhaps registered only imperfectly that the self-performance that brought her a popular audience was intrinsic to her critique of disciplinarity. On the lecture platform and in the narrative voice of her writing, the dual texts of self and science are virtually inseparable. Above all, Kingsley's work reflects a belief not only in the fundamental narrativity of science, but in its constructed nature and voice.[10] Demanding that her audiences attend to the person in the work, Kingsley entangles the stories of self and science to draw attention to a process of learning rather than remote pronouncements of science's definitive conclusions. Focusing on the all too human elements of the human sciences, she effectively avoids a charge of watered-down science in which oversimplifications are offered to satisfy an untutored audience, and instead represents the greater complexities of the practices of science. Through her presence, "her own peculiar way," Kingsley insisted that central to that practice, for good or ill, is its practitioner, a truth too often obscured in Olympian male discourse.

Kingsley's practice suggested that one could not begin to see the ways in which the observer is implicated until the European self, too, could be an object of scrutiny, itself destabilized and denaturalized. Her self-performance refuses the potentially possessive, all-comprehending eye of her audience just as her representations of West Africa disturb the possessive and certain gaze of imperial male narratives—either those of conquest or those indirectly serving conquest through science.[11] She closes the preface to *TWA* with a rather startling warning—and charge—to her readers: "Your superior culture-instincts may militate against your enjoying West Africa" (ix), she told them. Kingsley literally made a spectacle of herself in order to intervene in assumptions of her discipline. In self-performance, she denaturalized the proper woman and the proper scientist to focus on barriers to interpretation formed by an observer's too well-defined self with "superior-culture instincts"—an emphasis that had led Tylor to speculate that the quality of her work may have been enabled by a mysterious "genius and sympathy with the barbarian mind" (Green 7).

In this, Tylor, of course, reflects racial developmental theories grounding nineteenth-century ethnography and anthropology. Entwined with a self-justifying imperialist mission, disciplines concerned with cultural difference placed multifaceted difference on a linear scale ascending to Western European social organization and values. Kingsley too would announce herself "a Darwinian to the core," while also subscribing to a polygenist view of racial and sexual difference as one of kind not evolutionary degree.[12] While she would publicly rank different orders (the white man, the white woman, then all Africans), paradoxically the racism of essential difference also enabled her to insist on taking the West African on his or her own terms, and to rationalize her preference for a culture deemed savage. Tylor's view of her affinity with "the barbarian mind" reiterated her self-characterization in a letter of introduction to the Tylors: "I seem to have a mind so nearly akin to that of the savage that I can enter into his thoughts and fathom them" (1 October 1896). In this identification Kingsley appropriated, not possessive racial superiority, but nineteenth-century developmental theories that placed women and children closer than the white male to what was deemed the (particularly) black savage state.[13] The identification effectively "naturalized" the congeniality that West African social and spiritual systems held for her, and that Tylor had identified as "sympathy" and a "power of getting to the back of the negro mind." Tylor, however, also saw, incongruous to this schema, Kingsley's "genius," an unsettling acknowledgment producing the uncomfortable wonder threading through his praise of her work. Kingsley's work was certainly framed by a racist superstructure; once inside the frame, however, her interest focused on discerning the coherence and integrity of a cultural system. Kingsley defined her work, the study of fetish, rather simply as "the governing but underlying ideas of a man's [*sic*] life" (*TWA* 68). Her large view, theoretically as applicable to British "truths" as to any other culture's, became enacted in the spectacle of herself as British Woman, a produced cultural construction, no more natural than any other. Even more unsettling to careful readers such as Tylor, West African fetish ultimately appeared to Kingsley in many ways more coherent and congenial than Victorian Britain's fetish.[14]

On the lecture platform, Kingsley made both the observer and the observed "artifacts" worthy of study, and through her discursive tales foregrounded the interaction between them. Her work, particularly on the lecture circuit, would answer the question posed in an 1895 address by the retiring president of the American Association for the Advancement of Science, "The Aims of Anthropology": "But you will naturally ask, To what end this accumulating and collecting, this filling of museums with the art products of savages and the ghastly contents of charnel houses? Why write down their stupid stories and make notes of their obscene rites?" (Brinton 63). Certainly, paramount for Kingsley was the effort to disorient prevailing notions of "savage" and "ghastly," "stupid" and "obscene," and to question the methodological structure implied by "accumulating and collecting." Kingsley offered, not a catalog neatly labeled, but a multivoiced narrative of mutual misperceptions and perceptions, alternate scales of value, and, always, a sense of the provisional and partial nature of any "truths." She reoriented her audience to different angles of vision by including countless anecdotal sidesteps that center on exchanges between West Africans who are exceedingly smart about the Europeans they deal with and Europeans who remain remarkably stupid about their relations with West Africans.[15] In pointed contrast to theorists who assumed "the native" could not understand the origin or significance of customs and practices, Kingsley suggested a more realistic (and amusingly deflating) alternative: when seeking explanations, "[t]he usual answer is, 'It was the custom of our fathers,' but that always and only means, 'We don't intend to tell' " (*TWA* 477). In a practical application, Kingsley suggests challenges to a hierarchy of power in terms her audience can understand. England, "that nation of shopkeepers," for example, will find West Africans formidable in the Victorians' own terms: "[In Africa], young and old, men and women, regard trade as the great affair of life, [and] take to it as soon as they can toddle" (*TWA* 56). In Kingsley's view, on the matter of trade West Africans can hold their own, and it is up to the Englishman to prove that he is "an intelligent trader who knows the price of things" "(Lecture on West Africa" 267).

The European not only fails to see the West African apart from preconceptions, but also fails to see himself in the transaction. The two closely connected failures are brought together in Kingsley's insistence on learning to see the unfamiliar and on learning to see the familiar as unfamiliar—whether this is seeing "a racy maiden aunt" as scientist on the lecture platform, or seeing the presence of both the narrator and West Africans in her texts. Without remade vision, the scientist could be (or the audience could be gulled by) someone like "a German gentleman once who evolved a camel out of his inner consciousness. It was a wonderful thing; still, you know, it was not a good camel, only a thing which people personally unacquainted with camels could believe in" (*TWA* 10). Kingsley contrasts the vision produced by an "inner consciousness" with the gradual and receptive process of learning to see the forest: "As you get used to it, what seemed at first an inextricable tangle ceases to be so. . . . a whole world grows up out of the gloom before your eyes" (*TWA* 101); "The proudest day in my life was the day on which an old Fan hunter said to me—'Ah! you see' " (102). In Kingsley's lexicon, only relearned vision can ground, morally or intellectually, comprehension of the practices of West Africans: "At first you see nothing but a confused stupidity and crime; but when you get to see—well! . . . you see things worth seeing" (103). Kingsley, herself "a complicated criminal," as she had told Macmillan, and the West African, thought to embody "stupidity and crime," are indeed not what they seem, "but when you get to see—well!"

Kingsley's awareness of the complexities of vision and comprehension that must precede scientific interpretation strikes at the heart of the claim for mastery over another culture that was underwritten by professional suppositions of uninflected objectivity. Not surprisingly, reviews of her work during her lifetime are less interested than the obituary memoirs in puzzling over apparent incongruities between the style of the woman and the quality of her work and are more concerned with damping a disruptive, even alarming presence. Her most judicious critics find themselves unable to dismiss her work as unsound, but have instead to rely on distinguishing the higher and lower values of theory and praxis. One of *TWA*'s most significant reviews, Alfred Lyall's *Edinburgh Review* essay, places Kingsley in relation to F. B. Jevons and F. Max Müller. As a collector of materials, she is distinct from (and inferior to) the "philosophic savant" (Jevons and Müller) who "remains at home to receive what is brought to him . . . to classify, collate, and form his scientific inductions" (213). Kingsley was nonetheless pleased that the *Edinburgh Review* found her a force to contend with and was equanimous about Lyall's condescending (and gender-based) distinction, for much of her work, as the reviewer for *Folk-Lore* recognized, was a critique of armchair theorists:

> Miss Kingsley's repeated cautions to the anthropological student as to the reception and interpretation of evidence, the patience, the ingenuity, the tenacity of purpose, the open-mindedness required, and her warnings, none too emphatic, that no master-key will open all locks, are of a kind that ingenious theorists too often forget. ("Travels in West Africa" 163)

Kingsley, in fact, was an able—and cautious—theoretician who readily targeted the weaknesses of Frazer's "master-key," *The Golden Bough*, in her correspondence with Tylor,[16] and she well knew, as a later review of *West African Studies* pointed out, that in *TWA* "her observations [on the relation of witchcraft to religion] . . . have brought her athwart the theory of Sir Alfred Lyall and Professor Jevons" (Hartland 448).

Lyall uses both Kingsley's style and her method of organization to disqualify her work from the realm of higher science, hazarding that "she may peradventure have become unconsciously possessed by a jocose and humoristic fiend, whom in this Christian land she would do well to cast out" (214–15).[17] In his attempt to undo the threat of Kingsley's eclectic self-performance, Lyall reconstructs her as domestic worker, a scrubwoman to science, and makes the markers to a hierarchy of value clear: "From this curious and valuable description of primitive beliefs and customs in their natural state of entangled confusion we turn to the philosophic and well-ordered survey of their origin, interconnexion, and underlying psychology that is presented to us by Mr. Jevons" (224). Her work is both "curious and valuable," but Lyall cannot acknowledge the valuable disciplinary implications of its curiosity. In preferring an adaptation of the orderly classifications of natural science, he reflects an unwillingness to recognize, first, that one culture has indeed encountered rather than simply studied another, and second, that the encounter inevitably will be complex and untidy. On the lecture platform, Kingsley had the resources of her own physicality, her animated mannerisms, her incongruent appearance, and her discursive asides in the voice of the "jocose and humoristic fiend" to disorient the linking of authority and appropriateness. In person, Kingsley could use these resources to complicate schematic classifications as she liked. By framing her first book as a travel narrative, she established continuity between public performance and text through a genre that seemed most accommodating to her performative practice. Yet, in negotiations with Macmillan's editor, she found she would have to fight to include in her books the wealth of information that Lyall mistook for a "natural state of entangled confusion." As her letters indicate, her narrative "confusion" was carefully considered, and reflected a virtue in African storytelling: "Very few African stories bear on one subject alone, and they hardly ever stick to a point" (*TWA* 436). Kingsley contrasted her own work and writing with science's traditional, male model:

> These white men who make a theory first and then go hunting travellers' tales for facts to support the same may say what they please of the pleasure of the process. Give me the pleasure of getting a mass of facts and watching them. It is just like seeing a crystal build itself up. But it *is* slower I own. (To Alice Stopford Green, 27 March 1897, quoted in Birkett 173)

Her later book, *West African Studies,* includes a central portion of analysis reflecting her theorizing, the crystal that had grown, but she also insisted on retaining narrativity to show the crystal growing. She wrote Macmillan,

> The new book, though it will seem flippant enough and to spare when it is done, is heavy work for me. I am holding onto the main idea, round which it is written, by the scruff of its neck—but the selection of the facts that will bring that idea clearly out to the minds of people who do not know is hard work. (4 October 1897)

She reported one reader's response when she was preparing its opening section:

She always tells me . . . that I *ought not* to go on like that. Take myself seriously, etc. I really *am* always serious and 'duller than a great thaw' compared with the things I speak of, and I feel you really cannot understand W.A. unless you understand the steamboat. . . . this laughable stuff is in the thing—just as much as fetish is, etc.; and when Lyall and Mrs. G. and Guillemard and Strong and so on come along and expect me to stand on my head, all my innate vulgarity breaks out. (undated letter quoted in Gwynn 55)

Indiscriminate inclusiveness is here marked as a sign of class difference that separates her from the "white men who make a theory first and then go hunting . . . facts." Further, Kingsley's insistence that the steamboat "is in the thing" registers her acuity about adaptations of indigenous cultures to British imperialism. An outsider by class and gender, Kingsley uses her "innate vulgarity" to invert the delicacy of the "practices too disgusting to mention" school of ethnography. In this version the sense of delicacy or indelicacy shifts from the observed to the observer, and from science to the politicized self embedded in the science.[18]

Kingsley's agentless construction of a self in which "vulgarity breaks out" is a consistent and characteristic mode often borrowed by critics who see her as unable to control the "naturally" unconventional and her career as the fortuitous, if surprising, effect of that inability. Yet her success was clearly anything but natural. Kingsley carefully managed her career even before her December 1895 arrival from her second major trip to West Africa. She had proposed a book to Macmillan a full year before her return; a notation on her 18 December 1894 letter indicates "Miss Kingsley accepts 1/2 profits on her book of travels. Dec. 1894." In Liverpool, a Reuters news service reporter interviewed her: his story, a somewhat sensational summary of her "exploits" in pursuit of fish and fetish with a sprinkling of her highly quotable remarks, appeared as a news/feature in important British and American dailies. Subsequently, the *Spectator* took issue with some of those remarks, then printed her rejoinder, other readers' responses, and so on.[19] This initial "conversation" in the periodical press is an early measure of the ways in which she will be misunderstood: Townsend, the *Spectator* writer, errs in taking her comments to support his racist views, while in another *Spectator* piece responding to her corrections, she is chastised for appearing to defend cannibalism ("Negro Capacity"). The first month of her presence in the press established the ways in which she would stay in the public eye as she engendered and managed controversy. In the following weeks, she continued to elaborate and defend her views, thus adding the voice of the judicious (if controversial) interpreter of important scientific information to her status as a public event.[20]

Against her editor's advice, Kingsley's initial editorial column debates quickly gave way to consistent publishing in journals for an informed readership (she published twenty-seven articles in less than five years) and an exhausting lecture schedule. Guillemard was concerned that overexposure would cause the public to tire of her and that her unconventionality would damage her claims to serious science. She wrote Macmillan: "Dr. Guillemard has gone for me like a tiger for publishing articles. I am sorry if you similarly object to my having done so but my commercial instinct tells me that if I had not done so I should by now be forgotten by the fickle public" (6 July 1896). With her articles in periodicals like the *Liverpool Geographical Society* and the *Scottish Geographical Magazine,* "the fickle public" included the scientific community as well as the informed readership of *Cornhill* and the *National Review.* Kingsley saw no incompatibility of her "commercial instinct" with scholarly substance. At the same time that she was courting the G.P., she was writing Macmillan to urge *TWA*'s presentation as a scholarly work. "Personally I should like the book to be about the size and general get up of your Westermarck on Human Marriage" (1 May 1896).

Macmillan acquiesced in the book's "get up"; Guillemard's concerns about controlling Kingsley's public presence appeared, according to his lights, coherent with the stature of the work she intended. Indeed, the size and format of the published work were somewhat forbidding; its 743 pages, including five appendices and an index, belied its pose as a travel narrative, and its uncut pages—and stiff price of 21s.—announced a serious work. Nonetheless, the range of publications that reviewed *TWA* indicated that, after a year of Kingsley's omnipresence in periodicals and on the lecture circuit, its audience had become nearly everyone; published on 21 January 1897, it was in its fifth edition by June, and an abridged version appeared by the end of the year (Frank 230). *Punch* complained of its length, and other reviewers remarked on the onerous task of cutting so many pages; but, even if the book were more reviewed and talked about than read, the breadth of response validated Kingsley's commercial instincts. By the time that St. Loe Strachey reviewed her second book, *West African Studies,* for the *Spectator,* he could preface a passage from it with the unlikelihood, "In case anybody does not know how Miss Kingsley writes . . ." (169).[21] In many wholly respectable quarters, Kingsley had successfully forced a reconsideration of the appropriate demeanor of science and the scientist. That this was an issue—and an issue that she won—is reflected in the numbers of reviews that felt compelled to authenticate her: "Miss Kingsley is a true scientist," the *Church Quarterly* avowed ("West African Problems"); the *Folk-Lore* reviewer insisted she showed "the true scientific spirit" (Hartland); *Nature* found "much material of the greatest scientific importance" ("West African Fetish"); the *Dial* identified "a thoroughly scientific temper" (Stanley); and even the RGS's *Geographical Journal* grudgingly deemed her work "to possess permanent value" (Heawood).

Understandably, Kingsley found nothing threatening in the notion of popularizing science, if that meant bringing its processes under scrutiny, for she understood popularization as communicating the joys of process and the complex, provisional nature of discovery. Just as important, she represented the necessary but often serendipitous violations of methodology that conveyed the capacious vision and open mind required for good science. The qualities needed to practice science were also the qualities that Kingsley demanded of an audience that would watch a woman work. Kingsley's own experience could model an audience's response: "One by one I took my old ideas derived from books and thoughts based on imperfect knowledge and weighed them against the real life around me, and found them either worthless or wanting" (*TWA* 6). As a self-educated scholar, Kingsley applied a discerning critical judgment and demanded that audiences similarly evaluate authority independent of institutional credentials and "professional" demeanor. Despite her respect for the work of E. B. Tylor, "this greatest of Ethnologists" (*TWA* 435), Kingsley read and evaluated his work with great care. After hearing a paper that he had presented, she wrote to ask for a copy: "I want to read it and reread it for I am not smart on my intellectual legs, and like Mark Twain's horse frequently desire to lean up against a wall and think" (25 May 1898).

Kingsley's concern with reaching both the general and the scientific public reflected her understanding that science offered opportunities to challenge not only the demeanor of authority, but the methodology of authority; that those opportunities were themselves market-driven; and that they were not limited to the practice of science. In Kingsley's view, the General Public was an entity that could be named, marshaled, and empowered to resist an increasingly arcane and remote posture of professionalization across a broad spectrum.[22] Kingsley used herself as a prime "artifact" to communicate a conviction that the human sciences could be a space of meeting for marginalized voices, and that the empowered marginal voice could have significance extending considerably beyond a critique of the practice of science. Her debates on cultural and religious practices in West Africa, and on

British economic, political, and administrative policies, argued for interconnections of knowledge forming a larger project of reconfiguring cultural and gendered commitments. Kingsley's practice openly challenged the pure and isolate construction of scientific inquiry, and the imperial politics those constructions supported. Regrettably, the breadth of her interests has retrospectively been parceled out—largely to historical studies, area studies, genre studies, and feminist studies often focusing on Englishwomen in relation to the nineteenth-century British racism of empire. Splintering her interests, isolating what appear to be clear statements (yet necessarily choosing them from among disarmingly contradictory pieces of text) to bring together a satisfactorily coherent "whole," has not made Kingsley any less elusive today than she was to her contemporaries. "I foresee a liability to become diffuse," she said as the anecdotes in *TWA* multiplied. Readings of Kingsley's work that ignore that diffusion miss the destabilizing metadiscursive critique that her career embodies.

Kingsley was always concerned to make her information matter: by reconfiguring the processes of science, by revising the common view of West Africa, and by showing the fundamental importance of scientific understanding to the practical workings of the British Empire. As a "conscience of imperialism,"[23] Kingsley repeatedly argued in her lectures and in fiery periodical exchanges that scientific understanding of the coherence of West African practices and beliefs must be brought to bear on devising political and economic policies that would impose the least institutional apparatus and that would least interfere with West Africans. Correcting the missionaries' moralistic perception of one practice, she explained its practical significance, commenting to Tylor, "What a charming world that black world is—always so proper and so reasonable away down inside" (16 April 1898). Kingsley refused the categorization and hierarchical charting of practices that would override contextualization and instead pointed to a plurality of culture with insistence particularly upon the coherence of cultures of often disarming difference. Intelligent and respectful policy acknowledging in difference—and beyond difference—a world "so proper and so reasonable away down inside" would come about only under pressure from an informed public, a public with the tools for critically resisting the certainties of the authoritative voice.

Kingsley's management of her career—her impulse to take it to the public—owed much to her partisanship for free trade in West Africa, and to her hopes about the place of science within the common purview. However, just as her economic policies for West Africa swam against the tide of increased governmental administration, so too did her desire to make specialized knowledge accessible swim against the tide of increasingly determined institutional professionalization. Her career was overwhelmingly productive and successful, but, diminishing its potential to make a significant difference, unfortunately brief. Kingsley was in the public eye only a little over four years. She died in South Africa in June of 1900. Entering a discipline at a time when its methodologies were not yet "closed," Kingsley made substantial demands on its self-consciousness by forcing attention to the narrativity of science, and to the positioning of its narrator. With only a few short years of her public presence, however, the self-consciousness of the human sciences that she insisted upon was regrettably tabled for some time to come.

Mary H. Kingsley (1862–1900): Major Works:

Travels in West Africa, Congo Français, Corisco and Cameroons (1897)
West African Studies (1899)
Life in West Africa (1899)
The Story of West Africa (1900)
Notes on Sport and Travel (1900) [By George Henry Kingsley, with a memoir by his daughter Mary H. Kingsley]

Notes

My thanks to Sheila Sullivan for comments on drafts of this essay, to Martha Vicinus for guiding my thinking about the 1890s, to the University of Chicago and the NEH for financial support of research, and to the University of Alabama in Huntsville's Humanities Center for supporting travel to conferences.

1. In 1892–93, the RGS Council selected twenty-two "well-qualified ladies" for membership. The intensely debated policy was overturned; women were not again elected until 1913. Lord Curzon wrote to "contest *in toto* the general capability of women to contribute to scientific geographical knowledge" (*Times,* 31 May 1893). Gender also introduced class issues with "derogatory references to school teachers and governesses" (Birkett 219). See Middleton 11–16; Birkett 211–30.

2. Kingsley asked E. B. Tylor to sponsor her in the anthropological institute ("I am for a West Coaster fairly respectable & will not steal the other members' umbrellas or hats if I am allowed to join and pay my fee" [25 May 1898]). For the most part, Kingsley objected to women in professional societies, and found informal networks preferable. She wrote Alice Green: "Set yourself to gain personal power. . . . [T]he reins of power . . . are lying on the horse's neck; quietly get them into your hands and drive" (14 March 1900, quoted in Birkett 233).

3. Kingsley included this incident in an autobiographical essay ("In the Days of My Youth"). Other personal detail appears in her two-hundred-page preface to her father's *Notes on Sport and Travel* (1900). Both pieces stress the Kingsley connection, but Katherine Frank gives a more candid view of an unhappy childhood of isolation and social exclusion by the Kingsleys.

4. Kingsley first appeared in print (5 December 1895) angrily to rebut the *Daily Telegraph*'s New Woman label in their story of her arrival at Liverpool (3 December 1895).

5. A member of the audience recalled her appearance as "a bit of stagecraft designed to heighten her achievements" (E. Muriel Joy to Dorothy Middleton, 22 June 1966, quoted in Frank 258).

6. Highly regarding his work, Kingsley called Tylor her "great ju-ju" and initiated correspondence on returning to England.

7. The *Nation* termed it "racy . . . unconventional" ("Travels in West Africa"), although later cautioned that "the author . . . [falls] into colloquy, even into vulgarity, and almost profanity" ("West African Studies"); the *Bookman* noted "racy . . . even slangy English" (Dods); The *Illustrated London News* identified "a romping style" ("Notes on Books"); *Punch* saw "humour that bubbles over in all places" ("Our Booking-Office").

8. Frank summarizes some of her engagements (214–22, 234–40, 245–47, 252–58). Within eight weeks of her return to England, she wrote Macmillan detailing a punishing schedule: "I am going to Scotland for the reading of my paper at the RSGS then onto Glasgow . . . on the 12th I am to be in Liverpool for their Geographical and the Chamber of Commerce here have asked me for a paper. Professor Mahaffey . . . has also asked me to Dublin" (31 January 1896). By 1897 she had hired an agent (Frank 215).

9. The first article of the first issue of the African Society's journal (the society was founded in Kingsley's memory) was a tribute with reminiscences from friends and colleagues.

10. For foundational work, see Geertz; Clifford and Marcus. The latter considers the textuality of ethnography, but excludes a feminist perspective because, Clifford explains, "[feminist ethnography] has not produced either unconventional forms of writing or a developed reflection on ethnographic textuality as such" (21). In *Imperial Eyes* (1992), Mary Louise Pratt considers gendered positioning of narratives in relation to imperialist ideology. For specific discussion of gender and disciplinarity, see Moore.

11. The possessive, masculine view that feminizes the landscape has been widely commented on, e.g., Kolodny; Griffin. Ungendered cultural views appear in Said; Sternberger. Pratt's *Imperial Eyes* specifically considers the "eye" of scientific travel writing.

12. See Stocking's superb intellectual history for detailed analysis of currents of thought and their competing positions in the nineteenth-century development of the discipline.

13. In their survey of nineteenth-century intellectual currents concerning women, Helsinger et al. conclude, "[t]his equation of woman and black is one

of the most important features of the Woman Question" (2:91). For detail of the anthropological debate on woman's place in the hierarchy, see Stocking 187–237.

14. Examining fully the complexities of Kingsley's work in relation to race exceeds the scope of this essay. Kingsley publicly committed to political and scientific agendas of imperialism and the hierarchies of social Darwinism that embed racism. Readers who have found such precise lines inadequate to the representations and strategies of her texts have, however, tended to shape defensive arguments. In light of Kingsley's decentering strategies, approaches to her work that categorize it as racist, nonracist, or the apologetic "not-so-racist" may close down valuable inquiry into the incoherencies of acquiescence and resistance in the conflicted position of British women enabled by imperialism. This essay elaborates Kingsley's self-performance as one aspect of that incoherence: the specific nature of her gendered intervention in the professional discourse "denaturalized" the hierarchies underwriting nineteenth-century ethnography and imperialism at the same time that she committed to both.

15. In this essay, I use the broad term "West Africans" to discuss Kingsley's general perspective. Her books carefully specify indigenous peoples and cultures of the region.

16. Kingsley prefaced a ten-page letter to Tylor detailing Frazer's errors on animism and totemism, "My cap frills are vibrating with vexation" (9 April 1898). In *TWA,* she is dismissive: "I was particularly confident that from Mr. Frazer's book, *The Golden Bough,* I had got a semi-universal key to the underlying idea of native custom and belief. But I soon found this was very far from being the case" (435).

17. Virtually all reviewers commented on style before considering her positions. Many delighted in it, but others shared Lyall's distaste, expressing it with less avuncular preciosity. *Nature*'s angry reviewer felt that "hyperbole is frequently carried too far. . . . Serious students who, when they ask for facts, do not care to be offered a cryptic joke" ("Miss Kingsley's Travels"). The reviewer for *Science* allowed that "an easy flippancy of manner. . . carries you on, . . . [although] the writer is 'on very thin ice'. . . . The off-hand way in which some rather serious problems are treated is hardly fair" (Libbey).

18. Kingsley "vulgarized" herself in text by foregrounding breaches of conventional delicacy and an exaggerated horror of them. These incidents frequently involve mishaps with sex- and gender-marked clothing when she is with Europeans. Readings of these episodes as signs of a self-effacing and anxiety-ridden gender conservatism seem to miss the point.

19. Slightly different versions appear in the *Daily Telegraph* (3 Dec. 1895), the *New York Times* (2 December 1895), and the London *Times* ("Miss Kingsley's Travels," 2 December 1895). Her comments on "the nature of the West African" quickly elicited a *Spectator* article that misunderstood her (Townsend, 7 December 1895); she responded in the letters column (28 December 1895); the same issue ran an article responding to her letter ("Negro Capacity"). The *New York Times* then commented on the entire exchange ("African Character Studied,"

10 January 1896). The *Illustrated London News* ("Lady Traveler," 4 January 1896) profiled this new public figure (with photographs that she supplied). The RGS's *Geographical Journal* introduced her material: "Miss Kingsley has kindly sent, at short notice, the following notes on her recent journeys in West Africa" ("Miss Kingsley's Travels").

20. Less than three weeks after her return, Guillemard wrote, "I am quite a distinguished person here because I am a friend of Miss Kingsley. I enliven the dinner table with anecdotes about you. . . . Your book should run—I estimate—to about the 68th thousand, like Mrs. Henry Wood's *East Lynne* or Zola's *The Debacle*" (Macmillan Letters, 20 December 1895).

21. In the first week, 1,200 copies of *West African Studies* were sold (Frank 261).

22. Kingsley was acutely aware of ironies in distinguishing the professional from general public: "These literary and scientific institutions amuse me much. They . . . inform you they don't want science. . . . [but] 'something bright and amusing and magic lantern slides' " (to E. Sidney Harland, 25 March 1897, quoted in Birkett 203). Adapting to audiences from Oxbridge societies to Boys' Institutes in city slums, she profited from question-and-answer sessions to craft in text an implied dialogue considerably removed from the magisterial pronouncements that Lyall favored.

23. Kingsley's political activity was substantial. Historian Kenneth Dike Nworah argues that the self-named "Third Party" opposed the racist school that "advertised the inferiority and incapacity" of the West African, and the damaging philanthropic and missionary interests that would denationalize West Africa by eroding its traditional systems. The "Liverpool Sect" "was mainly apotheosized in the ideals of Mary Kingsley, John Holt, and E. D. Morel." In Nworah's view, the "small but perceptible sect . . . identified itself with the development of a true colonial conscience in Britain" (349–50). Although earlier free trade had meant the slave trade, Kingsley had faith in the process of knowledge (to which she contributed). Her advocacy reflected a utopian view of trade in which the skills of West Africans would ground an equality of interest with Europe and obviate the need for any "benevolent" control that would imbalance relations.

Works Cited

"African Character Studied." *New York Times*, 10 January 1896, 2.

Birkett, Dea. *Spinsters Abroad; Victorian Lady Explorers*. Oxford and New York: Basil Blackwell, 1989.

Brantlinger, Patrick. *Rule of Darkness; British Literature and Imperialism, 1830–1914*. Ithaca: Cornell UP, 1988.

Brinton, Daniel G. "The Aims of Anthropology." *Popular Science Monthly* 48 (November 1895): 59–72.

Bullen, Frank T. "Some Memories of Mary Kingsley." *Mainly about People*, 16 June 1900, 570.

Clifford, James, and George E. Marcus, eds. *Writing Culture: The Poetics and Politics of Ethnography.* Berkeley: U of California P, 1986.
Dods, Marcus. "West African Studies." *Bookman* 15 (March 1899): 179–80.
Fling, J. E. "Mary Kingsley—A Reassessment." *Journal of African History* 4, no.1 (1963): 105–26.
Frank, Katherine. *A Voyager Out: The Life of Mary Kingsley.* Boston: Houghton Mifflin, 1986.
Geertz, Clifford, *Works and Lives: The Anthropologist as Author.* Stanford: Stanford UP, 1988.
Green, Alice Stopford. "Mary Kingsley." *Journal of the African Society* 1 (1901): 1–16.
Griffin, Susan. *Woman and Nature.* New York: Harper & Row, 1978.
Gwynn, Stephen. *The Life of Mary Kingsley.* London: Macmillan, 1932.
Harland, E. Sidney. "West African Studies." *Folk-Lore* 10 (1899): 447–50.
Heawood, Edward. "Some New Books on Africa." *Geographical Journal* 13 (April 1899): 412–22.
Helsinger, Elizabeth K., Robin Lauterbach Sheets, and William Veeder. *The Woman Question: Society and Literature in Britain and America, 1837–1883.* 3 vols. New York: Garland, 1983.
Kemp, Dennis. "The Late Miss M.H. Kingsley." *London Quarterly Review* 94 (1900): 137–52.
Kingsley, George. *Notes on Sport and Travel, with a Memoir by His Daughter, Mary H. Kingsley.* London: Macmillan, 1900.
Kingsley, Mary H. "In the Days of My Youth; Chapter of Autobiography." *Mainly about People,* 20 May 1899, 468–69.
Kingsley, Mary H. "A Lecture on West Africa." *Cheltenham Ladies' College Magazine* 38 (Autumn 1898): 264–80.
Kingsley, Mary H. Letters to George Macmillan. Macmillan Papers. Correspondence. Manuscript Collection. British Library, London.
Kingsley, Mary H. Letters to Professor and Mrs. E. B. Tylor. Photocopies from originals in the possession of D. J. Holt. Manuscript Collection. Rhodes House Library, Oxford.
Kingsley, Mary H. "The Negro Future." *Spectator* 75 (28 December 1895): 930–31.
Kingsley, Mary H. *Travels in West Africa, Congo Français, Corisco and Cameroons.* London: Macmillan, 1897.
Kingsley, Mary H. *West African Studies.* London: Macmillan, 1899.
Kolodny, Annette. *The Lay of the Land: Metaphor as Experience and History in American Life and Letters.* Chapel Hill: U of North Carolina P, 1975.
"Lady Traveller in West Africa, A." *Illustrated London News* 108 (4 January 1896): 19.
Libbey, William. "Scientific Literature: Travels in West Africa." *Science,* n.s. 6, no. 139 (27 August 1897): 325–26.
Lyall, Alfred. "Origins and Interpretations of Primitive Religions." *Edinburgh Review* 186 (July 1897): 213–44.

Markham, Clements R. *The Fifty Years' Work of the Royal Geographical Society.* London: John Murray, 1881.
Middleton, Dorothy. *Victorian Lady Travellers.* Chicago: Academy Chicago, 1982.
Mill, Hugh Robert. *The Record of the Royal Geographical Society, 1830–1930.* London: Royal Geographical Society, 1930.
Mills, Sara. *Discourses of Difference: An Analysis of Women's Travel Writing and Colonialism.* London: Routledge, 1991.
"Miss Kingsley's Travels." *Times,* 2 December 1895, 6.
"Miss Kingsley's Travels in West Africa." *Geographical Journal* 7 (1896): 95–96.
"Miss Kingsley's Travels in West Africa." *Nature* 55 (4 March 1897): 416–17.
"Miss Mary Kingsley." *Times,* 6 June 1900, 8.
Moore, Henrietta L. *Feminism and Anthropology.* Minneapolis: U of Minnesota P, 1988.
"Negro Capacity—A Suggestion." *Spectator* 75 (28 December 1895): 927–28.
"Notes on Books." *Illustrated London News* 110 (6 February 1897): 185.
Nworah, Kenneth Dike. "The Liverpool 'Sect' and British West African Policy, 1895–1915." *Journal of the Society of African Affairs* 70 (July 1971): 222–35.
"Our Booking-Office." *Punch* 112 (20 February 1897): 88.
Pakenham, Thomas. *The Scramble for Africa: White Man's Conquest of the Dark Continent from 1876 to 1912.* New York: Avon Books, 1991.
Pratt, Mary Louise. *Imperial Eyes: Travel Writing and Transculturation.* London and New York: Routledge, 1992.
Robinson, Ronald, and John Gallagher with Alice Denny. *Africa and the Victorians.* New York: St. Martin's, 1961.
Said, Edward W. *Orientalism.* New York: Vintage Books, 1979.
Showalter, Elaine. *Sexual Anarchy: Gender and Culture at the Fin-de-Siecle.* New York: Viking Penguin, 1990.
Smith, Lucy Toulmin, and Mrs. Humphrey [*sic*] Ward. *Folk-Lore* 11 (1900): 348–50.
Stanley, Hiram M. "An English Woman in West Africa." *Dial* 22 (16 March 1897): 183–84.
Sternberger, Dolf. *Panorama of the Nineteenth Century.* Translated by Joachim Neugroschel, with an introduction by Erich Heller. New York: Urizen Books, 1977.
Stevenson, Catherine. "Female Anger and African Politics: The Case of Two Victorian 'Lady Travellers.'" *Turn-of-the-Century Women* 2 (Summer 1985): 7–17.
Stevenson, Catherine. *Victorian Women Travel Writers in Africa.* Boston: G. K. Hall, 1982.
Stocking, George W., Jr. *Victorian Anthropology.* New York: Free P, 1987.
Strachey, J. St. Loe. "Miss Mary Kingsley." *Spectator* 84 (16 June 1900): 836.
Strachey, J. St. Loe. "West African Studies." *Spectator* 82 (4 February 1899): 169–71.
Townsend, Meredith. "The Negro Future." *Spectator* 75 (7 December 1895): 815–17.

"Travels in West Africa." *Folk-Lore* 8 (1897): 162–65.
"Travels in West Africa." *Nation* 64 (1 April 1897): 249.
Walker, Bruce. "Travels in West Africa." *Athenaeum*, no. 3615 (6 February 1897): 173–76.
"West African Fetish." *Nature* 60 (13 July 1899): 243–44.
"West African Problems." *Church Quarterly* 49 (October 1899): 98–115.
"West African Studies." *Nation* 68 (23 March 1899): 228–29.

13

"Ape Ladies" and Cultural Politics
Dian Fossey and Biruté Galdikas

James Krasner

Dian Fossey and Biruté Galdikas had a hard act to follow. Louis Leakey had chosen them as the second and third of his famous "ape ladies"—women chosen to do long-term field studies of primates. Leakey's theory that women were better suited to such studies had, it seemed, already been borne out by Jane Goodall's well-respected and popular studies of chimpanzees. Goodall had gone on, with the help of *National Geographic* articles and television specials, to become an internationally famous scientist and the idol of every schoolchild. Now Fossey, studying mountain gorillas in Rwanda, and Galdikas, studying orangutans in Indonesia, were living in her shadow, hoping to make similar contributions and gain similar popular support.

Unhappily for them, times had changed. Donna Haraway has demonstrated how *National Geographic*'s popularization of the female primatologist in the nineteen sixties and seventies allowed Gulf Oil to portray an idealized reunification of western culture with nature while avoiding issues of colonialism, race, gender, and capitalist development. Jane Goodall, in particular, is portrayed as a "virgin-priestess in the temple of science and nature" (182) engaging in a "feminine gender-coded practice of identification and compassion" (184), redeeming twentieth-century scientific culture by evincing a maternal compassion for nature's animal children. As such, the portrayals of Goodall evoke one of what Marianna Torgovnick has identified as the two major western motifs of the primitive, in which the primitive world is "gentle, in tune with nature, paradisal" (Torgovnick 3), and contact between woman and ape creates "an oceanic sense of the oneness with nature" (181–82). Such an image of the relationship between nature and culture dovetailed nicely with American politics and popular culture in the sixties and early seventies.

But Fossey and Galdikas had to popularize their projects during the late seventies and eighties, during a resurgence of politically conservative conceptions of the relationship between first and third world, the role of women in science and culture, and the geographical semiotics of sexuality, race, and gender. During this period what Torgovinick describes as the "infernal primitive" becomes more culturally dominant than the paradisal. This more frightening construction of the primitive presents a third world landscape that is "violent [and] in need of control" (Torgovnick 3); it posits "associations between women . . . sexuality, and death" (162) and invokes images of insanity, human sacrifice, and cannibalism. Throughout the eighties, narratives of the infernal primitive are linked with colonialist nostalgia (Haraway 267). Films such as the *Indiana Jones* series, Busch's "Dark Continent" theme park, advertisements for Range Rover vehicles, Banana Republic clothing, Ralph Lauren's "Safari" cologne, all evoke a colonialist fantasy of western refinement surrounded by adventurous darkness. Africa is identified as a locus of disease and physical suffering in popular manifestations from the Live Aid concerts to rumors about the origin of AIDS. Narratives of Westerners "going native" emerge in films such as Coppola's remake of Conrad in *Apocalypse Now,* Theroux's *The Mosquito Coast,* and Milius's *Farewell to the King* and *Greystoke,* the latest, and most politically charged, remake of the Tarzan story. In all of these narratives gender roles are rigorously and traditionally defined, with women appearing either as demonic native seductresses or as virginal civilized maidens.

The *National Geographic* portrayals of Fossey and Galdikas also display images of the "infernal primitive." Generally speaking, Fossey is inscribed into a male narrative; she emerges as the dashing white adventurer descending into African darkness to fight heroically for the civilized values of scientific study and wildlife preservation. Galdikas, on the other hand, seems the naive white woman surrounded by primitive beasts and primitive men. These constructions are apparent in their own writings, particularly the *National Geographic* articles in which the infernal primitive becomes more pronounced as time goes by, and to an even greater extent in biographical and critical writings about them.

While neither became as famous as Goodall, both Fossey and Galdikas benefited from their exposure in *National Geographic*. Fossey achieved a considerable popular attention in the late seventies, appearing on the *Johnny Carson Show* and being featured in many documentaries. After her murder in 1985, a small Fossey industry developed, bringing tourists flooding to Rwanda and resulting in multiple biographies including Farley Mowat's *Woman in the Mists* (1987), Harold T. P. Hayes's *the Dark Romance of Dian Fossey* (1990), Sy Montgomery's *Walking with the Great Apes* (1991), Alex Shoumatoff's *African Madness* (1988), and the Hollywood film *Gorillas in the Mist* (1988). Galdikas remains in charge of a thriving, well-endowed orangutan study project. Her original study area has become a 250,000 hectare national park, complete with visitor's center, and she is in the unusual position of having upwardly mobile middle-class Americans fly to Borneo and pay large amounts of money to help her do fieldwork.

It is unfortunate that by making themselves visible to the popular media in the context of these narratives of the infernal primitive, these women scientists were forced to make themselves, rather than their studies, the focus of popular attention. Leakey's description of Goodall, Fossey, and Galdikas as "ape ladies," a term that seems more appropriate to a circus sideshow than a primatological field study, suggests that his choice of women rather than men had as much to do with his savvy at scientific popularization as with some protofeminist theory of scientific inquiry. Leakey no doubt understood that readers who would not be interested in evolutionary theory or animal behavior would be arrested by photographs of middle-class white women embracing apes. The popularization of these primatologists' studies involves, at all points, visual media: photographs, television, films. Rather than simply pictures of apes, Fossey and Galdikas's studies are illustrated by pictures of primatologists with apes; as at the sideshow, the blending of human and animal exists primarily as a spectacle. The taboos violated by white women entering animal worlds make for titillating reading, and the hypervisible image of woman and ape against a dark landscape invokes culturally conservative narratives about the relationship between nature, culture, and gender.

Dian Fossey wrote three articles for *National Geographic* in 1970, 1971, and 1981, and published her book-length autobiography *Gorillas in the Mist* in 1983. In all of these she is concerned with dispelling myths about the mountain gorillas, and with raising awareness (and money) in order to preserve them from extinction at the hands of poachers and Rwandan land developers. In the early articles, entitled "Making Friends with the Mountain Gorilla," and "More Years with the Mountain Gorilla," Fossey follows Goodall's example by focusing on her relationship with the sweet, funny, beautiful creatures she describes as "the gentlest of animals, and the shiest" ("More Years" 577). Happy family groups of gorillas frolic and play, tickling one another with flowers, and Fossey wonderingly enters this natural playground. But this paradisal world is replaced in her 1981 article "The Imperilled Mountain Gorilla" by a lurid and nightmarish landscape of brutality, social chaos, and doomed heroism. The article begins: "It was Digit, and he was gone. The mutilated body, head and hands hacked off for grisly trophies, lay limp in the brush like a bloody sack" (501). Fossey points out that Digit was one of many poacher-killed gorillas, whose numbers had been diminished "by half in just 20 years," and continues: "my aim [was] to balance research with the goal of saving the imperilled mountain gorillas that I was studying. . . . And now, through our sorrow, anger welled up—rage against the poachers who had committed this slaughter" (501). The gory image is meant to enlist the reader both in Fossey's rage and in her model of research balanced by preservation. A direct link can be traced between Fossey's growing interest in preservation over research and her use of the imagery of the infernal primitive. Throughout this article she focuses on violence and cruelty. "The poacher" is described as "bolstering his courage with hashish" and as killing gorillas "for killing's sake" (515). Photographs showing the mutilated bodies and body parts of gorillas abound. A detailed account is offered of Ian Redmond's near murder by a spear-throwing native. Surprisingly, Fossey also describes violence among the gorillas themselves. Infanticide, "murder" of one gorilla by another, gratuitous and ritualistic brutality by a young male gorilla against an older female, are all portrayed in vivid detail. Rather than just condemning poaching, Fossey seems to be aiming at an audience that wants to read about a savage and unlicensed jungle world, what Hammond and Jablow have characterized as the conventional literary Africa: "Death and violence emanate from the very atmosphere of such a land. . . . The 'blood soaked soil of Africa' serves as a basic premise for the literary reconstruction of African history" (137–38).

Haraway argues that Fossey went to her grave fighting for "her culture's dream" of an "African Eden" (Haraway 266–67). But five years before her death, Fossey had already replaced the African paradise with an African nightmare. Fossey's appeal to the reader is thus constructed along deeply conservative lines: Africa is a dark and dangerous place; gorillas are threatened because they live in Africa; white people must save the gorillas by carrying enlightened ideas about animal preservation into the darkness, thus saving the animals from the primitive natives and from themselves. At the end of *Gorillas in the Mist* Fossey represents wildlife preservation as a civilized practice, incomprehensible to the native Rwandans:

Foreigners cannot expect the average Rwandan living near the boundaries of the Parc des Volcans . . . to express concern about an endangered animal species living in those misted mountains. . . . American and European concepts of conservation, especially preservation of wildlife, are not relevant to African farmers. (239)

Haraway discusses how the portrayals of primatology in the sixties and early seventies are meant to unify first and third worlds, establishing a "touch across Difference" (134). But the establishment of Africa as a place of difference is essential to Fossey's rhetorical project; she must establish the distinction between nature and culture in order to invoke the dynamics of sympathy and inspire a paternalistic, fostering impulse.

Fossey's appeal was very successful. A photograph of Digit's mutilated body, accompanied by descriptions of the sinister Batwa poachers, was widely circulated after his death in 1977, and even appeared on the CBS evening news. As a result, the "Digit Fund" was established to help further support Fossey's work in Rwanda, and money poured into various gorilla aid organizations. It is not surprising, then, that in her extended autobiographical narrative, *Gorillas in the Mist,* Fossey should continue to represent Africa as an infernally primitive landscape and herself as a warrior against darkness. From the start Fossey places herself within a literary and cultural tradition of white male adventurers: "I spent many years longing to go to Africa, because of what that continent offered in its wilderness and great diversity of free-living animals. . . . [I] flew to the land of my dreams in September 1963" (1). Like Conrad's Marlow who yearned to find in the "blank space" of Africa "all the glories of exploration" (Conrad 10–11), Fossey presents Africa as a dreamworld of untainted wilderness, willing to "offer" a "free-living," uncivilized experience to any Westerner with enough gumption to "Come and find out" (Conrad 20). In the succeeding pages Fossey frequently hearkens back to literary antecedents like Conrad and Haggard. Like Allan Quatermain she finds herself in a magical African world, replete with monumental stone structures, marking the entrance to the wilderness:

Once within the thick bamboo, I felt a small bit of the magic of the wilderness. . . . From the bamboo belt the trail led through a cool rock tunnel, some five to six feet in width and about thirty feet in length. . . . The tunnel created a dramatic entrance into the world of the gorilla. It served as a passageway between civilization and the silent world of the forest. . . . (22)

Like Marlow she perceives herself as "alone" when surrounded by numerous Rwandan servants, but relishes the spiritual strength she gains from primitive "solitude":

The next day I felt a sense of panic while watching Alan [Root] fade into the foliage near the descending edge of the Kabara meadow. He was my last link with civilization as I had always known it . . . I clung to my tent pole simply to avoid running after him. . . As a pioneer I sometimes did endure loneliness, but I have reaped a tremendous satisfaction that followers will never be able to know. (7, 5)

As the book progresses, Fossey presents other formulaic elements of the imperial adventure narrative. At a camp Christmas party her African staff spontaneously places her in the ritual seat of honor, as they initiate a native celebration complete with traditional drumming (166). She fights off black magic, a witch doll carved in her own image, with jovial common sense (34–35). There are repeated adventures, in which she is imprisoned by the army, stalked by poachers, nearly attacked by gorillas, all of which she narrowly escapes. She even falls into a pit trap, and must scramble out before her captors find her (28). In short, Fossey becomes the great white hunter, idolized by native servants, misunderstood by European neophytes, wizened and scarred by the rigors of life in the jungle.

After Fossey's murder in 1985, the narrative of western explorer "going native" is taken up by popular biographers, who expand and pathologize her story along the primitivist lines she established. In these portrayals Fossey moves from battling with darkness to succumbing to it; she becomes Kurtz rather than Marlow, driven mad by her mission, making herself a demonic god to the natives, and consorting sexually and spiritually with the powers of African darkness. In Alex Shoumatoff's *African Madness* (1988), Fossey's "Africanization" is described, like Kurtz's, as the transposition of the colonial hierarchy onto a primitivist model of tribal authority and magical power. Shoumatoff quotes Kelly Stewart, one of Fossey's research assistants, describing her punishment of poachers:

She would torture them. She would whip their balls with stinging nettles, spit on them, kick them, put on masks and curse them, stuff sleeping pills down their throats. She said she hated doing it . . . but she got into it and liked to do it and felt guilty that she did so. . . . She reduced them to quivering, quaking packages of fear, little guys in rags rolling on the ground and foaming at the mouth. (Shoumatoff 28)

By beating the poachers at their own game, using their superstitions against them, Sy Montgomery suggests in *Walking with the Great Apes*, Fossey plays the typical role of Westerner gone native, forcing the inferior race to grovel in rags before her.

> She painted hexes, cast spells, pronounced curses. . . . "I am the Goddess of the Mountain," she would hiss in KiSwahili, "and I will avenge you for killing my children." . . . Faking the supernatural is an old colonial trick; to convince Africans of their mythical powers, early explorers would do everything from discharging firearms to pulling out their false teeth. Dian's ventures into witchcraft were more than a trick; they became a sacred rite. (Montgomery 221)

Montgomery places Fossey in a line of great colonial explorers, rather than scientists, as Fossey had already done in *Gorillas in the Mist*. The encomia to Fossey, like those to Kurtz, both admit her abuses and assert that she was a "very remarkable person" (Conrad 29). Interestingly, Lutz and Collins describe the *National Geographic* photographer as fulfilling the role Fossey wanted to construct for herself:

> The *Geographic* photographer has always been . . . the great hunter/adventurer . . . free to roam the globe in search of visual treasure, flamboyantly virile in . . . his bravery in entering the dangerous realms at the ends of the earth, in continents still dark for most of his audience. (184–85)

It is not surprising, then, that Fossey's vision of herself blended so well with the magazine's photographic mythmaking. Fossey stands in the foreground wearing safari clothing with Mt. Visoke rising majestically behind her and a thin line of African porters carrying crates on their heads winding into the distance ("Making Friends" 80). Fossey sits at a camp table, reading intently, while an African servant raises an axe to one of the towering jungle trees ("Making Friends" 56).

But Fossey's photographers could not fully accept her as a fellow male imperialist adventurer; they expected her to address the camera as a woman. Bob Campbell, the *National Geographic* photographer responsible for both still photographs and films of her interactions with Digit, complained about her failure to accept her feminine role:

> Off the mountains . . . with rest, makeup, and feminine clothing, Dian could transform herself into a startlingly attractive woman. But in Rwanda she did not often make the change. I had great difficulty with her looks in the field—in many situations where there was no possibility of preparations, she looked terrible. (Campbell quoted in Hayes 217)

As a result, pictures of Fossey become notably fewer in later articles, and she appears in ridiculous or unflattering poses. In one she is seen imitating Digit as he scratches his head and belly, her mouth screwed up into an idiotic grin ("More Years" 578). In another she appears sweating, dirty, and rumpled, huddling down among a group of wet gorillas ("Imperilled" 504). Haraway makes the point that all female primatologists are linked with male photographers who mediate between them and their audience (150). Here that mediation amounts to a gender-defined deconstruction of Fossey's self-created image; rather than continuing to portray Fossey as the white hunter, Campbell presents her as the white woman going to seed by failing to live up to western standards of beauty.

Her popular biographers also reinterpret Fossey's persona using gender categories. Because she is a woman, she is read alternately as doomed adventurer and mystical jungle queen—Kurtz and La, Quatermain and She. The combination of sexuality with sexual torture, madness, witchcraft, and violent authority suggests the sort of demonic, matriarchal character Torgovnick describes in the Tarzan novels: "Part primitive ruler, part S & M queen, La and characters like her adorn several of the Tarzan covers, sometimes caught in the moment of raising their long knives, sometimes surrounded by dead men or animals or skulls" (64–65). Like La, Fossey is a sadistic imperial white ruler surrounded by black servants and beasts. Also like La, she tortures and hexes her enemies in sexually sadistic ways, and is pictured among decapitated and mutilated bodies and skeletons. In her 1981 *National Geographic* article a poacher is shown tied to a stake; at first glance he seems to be the victim of human sacrifice. Her biographers also portray her sexuality as both predatory and insatiable. In Harold T. P. Hayes's *The Dark Romance of Dian Fossey* she is an ugly, domineering sybarite, pursuing photographers and research assistants alike (223). "We could have shacked up together and had a hell of a good time," she reportedly tells John Alexander. When he demures because he is in love with another woman, she retorts, "what makes you think sex has anything to do with love?" (Hayes 93). Photographer Bob Campbell "had little sexual interest in Dian Fossey" because of her ugliness, but Fossey pursued him relentlessly "until on one occasion Dian decided to make love to me, and I did not resist" (Hayes 223). Montgomery describes how she "subsisted for months on great crates of pornography she had friends ship to her, and she kept a vibrator to satisfy her large appetite for sex" (154).

Perhaps the most common characterization of Fossey by later biographers is that she becomes like an animal. Montgomery writes that "Dian was like a silverback [an adult male gorilla] who raided other families for its mates" (154). Alex Shoumatoff quotes Ian Redmond, another research assistant, to the same effect: "Dian as an individual was in many ways like the gorillas, . . . in that if you are easily put off by bluff charges, screaming and shouting, then you probably think that the gorillas are monsters"(29). After describing how Fossey took a mink stole along on her first trip to the jungle, Montgomery offers the following jarring anecdote:

> By 1977 . . . Dian had been living in the mountains for ten years. She was forty-five. As [she and two graduate students] sat in a booth in a hotel restaurant waiting for their order to arrive, Dian began licking pats of butter off the cardboard backings and sucking sugar out of the paper packets. (133)

Seen through the eyes of graduate students eager to emulate a respected scientist, Fossey emerges as a social outcast—an irrational eccentric, a street person, an animal. While a certain amount of antisocial eccentricity is expected from male scientists (Einstein forgot to put his pants on), Fossey emerges as an unacceptable degeneration of feminine social refinements into bestial physicality. One of the crueler truths about the way in which Fossey's story has played itself out in popular media is that her status as a scientist has been eclipsed by her role as heroic madwoman confronting the primitive.

In one particular *National Geographic* photograph Fossey leans, smiling, over an array of darkened gorilla bones and skulls ("More Years" 576). This picture demonstrates how her scientific methods became imagistically translated into enactments of the narrative of infernal primitivism. While Haraway sees Goodall as a representative of western science, Fossey's very scientific activities (her isolation, her physical deprivations, her administrative authoritarianism, her physical interaction with the gorillas) become representations of antiscientific, irrational behavior, evocative of witchcraft, obsessiveness, madness.

Where Fossey constructs her *National Geographic* stories so as to appeal to the reader's desire for adventure, invoking the myth of Westerner going native, Biruté Galdikas appeals to her reader by emphasizing her vulnerability to, rather than her struggle with, a violent and destructively sexual primitive world. She portrays herself in a more traditional female role than does Fossey: the white woman at risk from threatening beasts in a threatening landscape. Galdikas's stance toward the apes is also more traditional in that it is explicitly maternal, but maternal and sexual roles undergo a transgressive blending in the infernal primitive world, making Galdikas once again a victim. Haraway discusses how the Galdikas stories contain "ambiguously innocent and transgressive reproductive polity" between women and animals, but interprets these sexual interactions as part of a master narrative of the "quest for origin" that leads to the healing of "original sin" (Haraway 142–43). By emphasizing Galdikas's healing, maternal role, Haraway underplays the extent to which Galdikas's articles in *National Geographic* (1975 and 1980) and *International Wildlife* (1990) present her body as the object of both desire and violence.

Galdikas offers an extremely embodied portrayal of herself; she emerges both as the suffering, vulnerable, and potentially violated body and as the nurturing and reproducing multi-species maternal body. In both of these representations Galdikas offers her reader a landscape of suffering. Whether it be primatologists or infant orangutans who suffer, the dynamics of sympathy emerge much more powerfully and directly than they do in Fossey's writings. The reader is moved to help Galdikas and, by helping her, to help her infant orangutans.

In order to offer as vivid a portrayal of weakness in need of strength as she can, Galdikas frequently casts herself in dramatic tableaux that play on traditional constructions of feminine vulnerability. In her description of a near attack by a male orangutan, for example, she cowers behind her protective, machete-wielding husband, clutching her (orangutan) child to her breast:

> Sugito was terrified. He slid down the last few feet of the tree as though it were a greased pole and leaped into my arms. T.P. was right behind him.
>
> I stood up, panic-stricken, and backed away. Had Rod not been there, I think I would have run for my life. And, doubtless, T.P. would have chased me. But Rod calmly stood up with machete in hand and, looking directly at T.P., sliced through a small sapling just in front of him.
>
> At this T.P. stopped; . . . This was the closest that a wild orangutan ever came to attacking us. ("Orangutans" 460)

Earlier in the same article, Galdikas describes her first meeting with T.P.; the "huge" "adult male" faces the "pale-faced primatologist with large black sunglasses, clutching an enormous bag full of dirty laundry" (447). Galdikas's emphasis on her whiteness, her fear, and particularly her semiotic positioning as a traditional wife doing laundry and holding babies, belies her position as the leader of a scientific field study. But they do make for easily imaged scenarios, allowing the *National Geographic* reader to dramatically contrast the familiar middle-class white world with the primitive landscape.

The sexual vulnerability and visibility of the white woman in the primitive landscape also emerges in Galdikas's portrayals of herself. Galdikas spends a great deal of time describing the physical privations she undergoes; two out of her three articles start with lengthy descriptions of wounds. Rather than presenting herself in the tradition of the male quester, undergoing hardship for glory, Galdikas presents wounding in sexually provocative terms, inviting the reader to look at her violable body:

> I had left the forest and was limping back to camp in the rain. My left leg was soaked with blood from a wound made by my machete, which had slipped as I cut a vine. . . . The swamps, swollen by rains, were waist deep and impassable. Leeches were everywhere. Bloated with our blood, they fell out of our socks, dropped off our necks, and even squirmed out of our underwear. ("Orangutans" 444, 446)
>
> We were a sorry sight. I was bleeding from leech bites, caked with mud and soaked through with rain, and so was Rod Brindamour, then my husband. I was also in terrible pain. I had sat on a log oozing toxic sap, and a large area of my behind looked burned, like an overly toasted marshmallow. ("My Life" 34)

The particular detail about leeches in the underwear appears in two of the articles, and is taken up by Sy Montgomery, who offers an extended description of leeches that "come toward you from every direction like heat-seeking missiles; they loop forward like inchworms, standing upright and waving their mouths in the air, sensing your warmth" (165). The leeches seem both sexual and violent, and they fulfill the primitivist convention of a landscape full of "pullulating life which feeds on itself—devouring or being devoured" (Hammond and Jablow 137), while at the same time giving the reader imaginative access to Galdikas's underwear. Similarly, while Galdikas draws the reader's attention to her "behind" to emphasize her suffering, the invitation to look at an erotically charged part of her body turns the emphasis of the passage from suffering to unveiling. In both passages the reader is encouraged to look—to perceive Galdikas's private body as something visibly accessible—and in both cases the look moves toward a grotesquely erotic image.

Her mothering of the infant rehabilitant orangutans also has strong sexual overtones. Galdikas describes her "monster babies," as she calls them, giving her "French kisses" and becoming sexually jealous of human males. The following description of her care for the infant orangutan Sugito offers a combination of maternal and sexual images:

> Determined to give him as normal an upbringing as possible, I allowed Sugito to cling to me. I had little choice. Even shifting him from one part of my body to another involved much fighting and howling. Changing clothes became a major undertaking, with Sugito screeching and clutching at whatever was coming off. He slept curled up next to me and would not abandon me even when I bathed in the river. ("Orangutans" 451)

The description focuses on changing clothes, sleeping, and bathing, rather than on eating, carrying things, or engaging in fieldwork; it thus highlights the three practical daily activities most fraught with sexual suggestiveness and demonstrates how the maternal and the sexual become blended in Galdikas's portrayal of her life in the field.

Haraway discusses how one photograph in the initial *National Geographic* story "emphasizes [Galdikas's] breasts" in order to portray her as a "maternal body" (148). As Lutz and Collins have pointed out, *National Geographic* has, throughout its history, eroticized the female photographic subject (175). The fact that white women are rarely so eroticized suggests that Galdikas, in her interaction with the apes, is coded as closer to nature and thus more erotically available. Her poses seem designed to flatter her figure; many of them seem like conventional model poses. She generally wears low-cut tops with only a few buttons fastened. In many pictures the orangutans are pulling at her clothes, thus exposing more of her body ("Orangutans," cover, 472; "Living" 836, 839, 841, 851). In one picture a half-grown orangutan is shown looking up a female visitor's shirt because, the caption tells us "the author hides treats under her clothes" ("Living" 851). Several of the pictures were torn out of the library edition I was using.

In their discussion of *National Geographic*'s aestheticization and commodification of women's bodies, Lutz and Collins describe how "glossy images of women led the way in selling the third world to travellers" (177). In Galdikas's case this means selling the preservation of the third world. While Fossey's reader is moved by outrage into a fostering impulse, Galdikas's is moved by a paternalistic sympathy tinged with eroticism. Galdikas portrays herself as vulnerable; this vulnerability manifests itself both as suffering and as availability to the reader's voyeuristic gaze.

The voyeurism and gender determination set up in the *National Geographic* articles are played out by other popularizers of Galdikas's story. In more recent portrayals, the erotic rivalry between the animal gaze and the human gaze has intensified, and its underlying racial aspect has emerged. Like the leeches, the orangutans represent sexual, as well as physical, danger:

> When Biruté first came to Ranjung Puting, the rubber trappers told her that male orangutans will also rape human women. . . . Biruté knew that many mammals, from dogs to cougars, become excited by the scent of menstruating human females, but she dismissed the legends about orangutans. She would walk through the forest with blood soaking through her jeans each month for Kotex and tampons were not available in the nearby towns.

> Years later a wild adult male orangutan came into her camp and raped a female Indonesian cook. (Montgomery 174–75)

The strong implication here is that Galdikas barely escaped rape herself. Montgomery proceeds to describe orangutan mating practices, noting that "They make love as humans do, belly to belly" (174), moving quickly into a discussion of Galdikas's divorce from her first (white) husband, Rod Brindamour, and subsequent marriage to Pak Bohap bin Jalan, a Dayak, who is described as follows:

> Pak Bohap's forest skills are unparalleled. He can run barefoot through swamps. In seconds, while holding a lit cigarette in one hand, he can climb a tree and triangulate precisely the best spot from which to view an orangutan from the ground. He can tell you, from looking at bent twigs, which animal passed by here and how long ago. . . . he can pinpoint the source of any forest sound . . . with uncanny accuracy. (178–79)

Montgomery's gliding from orangutan rape, to orangutan mating, to the human mating of Galdikas and Pak Bohap in the space of a few pages points to the sexual and racial linkages Haraway discusses (Haraway 152–56). The threat of white women mating with men of color, coded as white women mating with animals (Haraway 153), is thus both revealed and realized in Montgomery's portrayal of Galdikas's marriage.

In portrayals of both Fossey and Galdikas, then, traditional fears for white women in colonial settings emerge in fairly explicit and traditional forms. Haraway makes the important point that the early seventies portrayals of female primatologists avoid difficult issues such as race and gender by positing an ahistorical model:

> The stories are about modes of *communication*, not history. "Communication" . . . is about boundary crossing, about the drama of touch across Difference, but not about the finite, difference-laden worlds of history. The duality nature/culture works well within this frame of communication because both of its terms are allochronic, i.e., existing in time outside the contentious, coeval time of history, with all its differences and uncertainty of understanding. (149)

Eighties portrayals are also allochronic, but rather than portraying the fundamental and timeless unity of nature and culture, people and animals, first and third world, they portray a fundamental and timeless difference between these categories. The touch across Difference becomes a violation of essential, and socially crucial, norms rather than an expression of a socially redemptive communication. Haraway suggests that by introducing issues of race, gender, and politics, by portraying rather than avoiding "contentious . . . differences and uncertainty of understanding," one necessarily historicizes a representation. In the eighties portrayals, however, we see that one can portray a difference-laden world without portraying the difference-laden world of history.

Often, Fossey's and Galdikas's status as field scientists does not square with their roles in the primitivist narrative. We have seen how representations of both Fossey and Galdikas play down their scientific role; their articles and books are structured biographically; the narrative climaxes of the articles consist of dramatic encounters between women and apes rather than scientific discoveries. Popular works about them tend to carry this even farther, focusing on moments when the primatologist rejects scientific study in favor of a "nonscientific" engagement with the apes. When, in Farley Mowat's *Woman in the Mists,* Fossey saves a baby gorilla kidnapped by a Cologne Zoo, "Field studies were forgotten as she nursed the young gorilla day and night . . . even sleeping with it in her own bed" (75). Sy Montgomery says of Galdikas that "the bottom line, as with Jane and Dian, is that science is not her top priority. Science was the reason she first went into the field, but science is no longer what keeps her there" (189).

The contribution of Leakey's "ape ladies" to a tradition of women scientists' self-presentation must be read as very mixed. Within these conservative cultural narratives women scientists are required to define themselves through or against cultural stereotypes of the feminine even as they are engaged in scientific study. Their visible embodiment in *National Geographic* emphasizes their proximity to a physical nature rather than an intellectual culture. In *Reading National Geographic,* Lutz and Collins have pointed out how the photographic semiology of *National Geographic* is based on the contrast between only two worlds, coded culture and nature or modern and traditional (110–14). By adopting this two-world model and representing themselves at the visible border between these two worlds, Fossey and Galdikas both place themselves in a culturally subordinate position, and become ideal subjects for the magazine's style of scientific popularization.

Dian Fossey (1932–1985): Major Works

"Making Friends with the Mountain Gorilla," *National Geographic* (1970)
"More Years with Mountain Gorillas," *National Geographic* (1971)
"The Imperilled Mountain Gorilla," *National Geographic* (1981)
Gorillas in the Mist (1983)

Biruté Galdikas (b. 1946): Major Works

"Orangutans, Indonesia's People of the Forest," *National Geographic* (1975)
"Living with the Great Orange Apes," *National Geographic* (1980)
"My Life with Orangutans," *International Wildlife* (1990)
"Birth Spacing Patterns in Humans and Apes," *American Journal of Physical Anthropology* (1990)

Works Cited

Conrad, Joseph. *Heart of Darkness.* New York: Bantam, 1989.
Fossey, Dian. *Gorillas in the Mist.* Boston: Houghton Mifflin, 1983.
Fossey, Dian. "The Imperilled Mountain Gorilla." *National Geographic* 159 (1981): 501–23.
Fossey, Dian. "Making Friends with the Mountain Gorilla." *National Geographic* 137 (1970): 48–68.
Fossey, Dian. "More Years with Mountain Gorillas." *National Geographic* 140 (1971): 574–85.
Galdikas, Biruté. "Living with the Great Orange Apes." *National Geographic* 157 (1980): 830–53.
Galdikas, Biruté. "My Life with Orangutans." *International Wildlife* 20 (1990): 34–41.
Galdikas, Biruté. "Orangutans, Indonesia's People of the Forest." *National Geographic* 148 (1975): 444–73.
Hammond, Dorothy, and Alta Jablow. *The Myth of Africa.* New York: Library of Social Science. 1977.
Haraway, Donna. *Primate Visions: Gender, Race, and Nature in the World of Modern Science.* New York: Routledge, 1989.
Hayes, Harold T.P. *The Dark Romance of Dian Fossey.* New York: Simon & Schuster, 1990.
Lutz, Catherine A., and Jane L. Collins. *Reading National Geographic.* Chicago: U of Chicago P, 1993.
Montgomery, Sy. *Walking with the Great Apes.* Boston: Houghton Mifflin, 1991.
Mowat, Farley. *Woman in the Mists.* New York: Warner, 1987.
Shoumatoff, Alex. *African Madness.* New York: Knopf, 1988.
Torgovnick, Marianna. *Gone Primitive: Savage Intellects, Modern Lives.* Chicago: U of Chicago P, 1991.

PART 8

THE TRADITION CONTINUES

14

Interview with Diane Ackerman, 18 July 1994

Barbara T. Gates and Ann B. (Rusty) Shteir

Rusty: Have you thought of yourself in connection with a tradition of women writing about science?

Diane: No, but I have thought of myself in terms of a tradition of women—and men—writers. For instance, when I think of somebody like Colette, I feel a sense of community. Colette was writing about nature inadvertently, she wasn't setting out formally to do it, but she had such a keen, such an affectionate curiosity about the world around her. She wrote about her gardens, and about nature and animals, with the kind of tact and attentiveness that naturalists do. So I don't feel limited by writers who are in a certain category. I suppose John Donne or Lewis Thomas influenced me as much as anyone. Virginia Woolf and Colette and various other writers, just because of the way that they perceived nature, also influenced me.

Rusty: To carry on with women writers who matter to you, I'm very taken by your poem about Sor Ines Juana de la Cruz.

Diane: Now there's a woman who really defied categories. She was a natural scientist, a composer, a poet. She was a

Figure 14.1. Diane Ackerman. Photograph© Jill Krementz.

very courageous and extraordinary woman who did wonderful experiments of the sort that Newton was doing—and at the time that Newton was doing them! So yes, I do find her a kindred spirit as well, because she had such a metaphysical mind. To go back to Donne, she was as interested in the human side of any question as she was in the scientific side of any question. Of course, I respond to her.

Rusty: Is that a form of self-definition for you?

Diane: In my case, when I was in graduate school I did my degrees in English and comparative literature—but, I was always poaching in the sciences. "Physics for Poets," I remember, was one course that I took. I never could quite figure out if I wanted to be in the arts or in the sciences. I didn't believe, and don't believe now, that the universe is knowable from only one perspective.

Barbara: I wonder if you do something similar in terms of traditional genres. You write the identical sentences or phrases in your prose and your poetry, and your prose and poetry talk to each other. Would you comment on that?

Diane: What happens is that I end up writing several of my books at the same time. My muse is sufficiently miscellaneous that I might be, let's say, in the Antarctic writing an essay about that astonishing landscape and penguins for a book of essays. At the same time, I'll be noticing fascinating sensory spectacles, and that will go into the *Senses* book, and at the same time there'll be some creative and emotional overspill that will go into a suite of poems set in the Antarctic. All of that seems to be happening simultaneously, but in different places in my mind. I think of it as having a captain's desk with many drawers. When one drawer is open, it's completely open (the others are closed), and it has my full attention. And when that's closed, it's closed, and I open up the next one. But the drawers contain essay, personal essay, scholarly essay, poetry—whatever it's going to be—all at once.

Even when I'm writing prose—and most of what I write these days is prose—the source of my creativity really is in poetry. It takes me longer to write a poem than it will take to write twenty times as much prose, fifty times as much prose—there is something denser and more fulfilling for me about writing one page of poetry that somehow captures exactly the emotion that sponsored it than writing fifty pages that also perhaps capture the emotion.

Rusty: Part of the history of women's writing is that women write miscellanies.

Diane: That's a very good point. It might well be that women are more attracted to that kind of work because they're more attracted to the integration of miscellaneous aspects of the world.

We should ask ourselves whether a man's relationship with his body is the same as a woman's relationship with hers. Is a man's sense of his body in nature the same as a woman's in nature? When women write about nature, is there in fact more of a sense of nurturing and more of a miscellaneous sense of integrating all the different threads, or not?

Rusty: Do you have any sense of that in terms of subject areas? I'm thinking of your "Defraction" poem for Carl Sagan, where you make the distinction between mathematics and other areas.

Diane: Bear in mind that I wrote that when I was about twenty-two!

In the Carl Sagan poem I really saw a difference between the mind-set of the scientist and the mind-set of the poet, and I envied the mind-set of the scientist because I thought at that time that he had a better grip on the natural world. Now I've learned that isn't necessarily true.

Barbara: But you still find the poetry in science. For example, in your essay in the *New York Times* (29 May 1994), "Mute Dancers: How to Watch a Hummingbird," you begin a short prose piece with a line that also enfolds your poem "The Dark Night of the Hummingbird":* "A lot of hummingbirds die in their sleep." The line perfectly arrests the reader with the tragedy of these beautiful, fast-living creatures, which, when they slow down, may slow down to death. Both the poetic prose piece and the poem are little jewels, like the birds themselves, and scientifically accurate, built on fact.

*Poem reprinted at the end of the interview

Diane: That's a very good example of how obsessive a poet can be about an image that has meaning. Some images nag at you, they stay with you, and you don't know why exactly, but they bother you in some profound, human way.

So I read, in *Science News* or *New Science* or one of the journals that I subscribe to, about the plight of hummingbirds, and how they very often die in their sleep. This registered powerfully because I was thinking that they live at an intense passionate pace, as do artists, as do lots of people. The cheapest way to say this is you pay for your thrills, but the more complicated way to say it, I tried in the poem. I also wanted to say the poem in a form that would capture some of the iridescence in language of the hummingbird. I know it's not a direct parallel . . . how can you take words and convert them into what will seem visual in that way? How can you make the form "iridesce"?

But then, as I was working on a prose book, I thought: "I just need to think about this more, and I need to think about it in a place where I have elbow room and can make it a little more colloquial and bring in anecdotes that girlfriends have told me about their experiences with hummingbirds." So you're right. I then decided to do the *Times* piece, which turned out to be a kind of long prose poem with science in it and miscellaneous other things as well.

Rusty: Historically, women writing popular science often chose forms that were colloquial, informal. They used conversations or letters.

Diane: Why would you want to take all aspects of life and separate them, formalize them? I'm sure this is anathema to people who believe you should compartmentalize, but that's essentially why I'm not a scientist. I'm interested in the truths of science, and I spend a lot of my life discovering some of those truths myself. If you're working from nature and observing animals, you see things that haven't been seen before. You record things that haven't been recorded before.

I'm interested in all of the universe. It interests me in detail and it interests me as an interrelated whole. The way you get Nobel Prizes may be to put on blinkers and look at one tiny aspect of the world until you understand everything about it, but that doesn't appeal to me much. I would prefer to see the forest for the trees. No, I would prefer to see the forest *and* the trees.

Barbara: You have titled two of your major works *A Natural History of the Senses* and *A Natural History of Love*. Why did you use the words "natural history"?

Diane: For me, a natural history is an umbrella phrase that means looking at something from as many perspectives as possible—as nature, as history, as sociology, as psychology, as personal experience. It's hard to figure out what to call such books because they are long contemplations. The best book about the senses or about love would take your entire life and include all of consciousness, and who can do that? So you have to select. It means that you're going to take something many-faceted and choose a few of the facets that you think will reveal the most sparkle and illumination. But the whole enterprise is not really a history, it's not really a scholarly work, it's not really a scientific treatise. The best phrase I could figure out was a catchall phrase, a "natural history."

Rusty: Did you have ideal readers in mind for your two "natural history" books?

Diane: I never worry about that. I get letters from people who share their memories and experiences with me. Occasionally somebody will say to me, "If you wondered whom you were writing these for, you're writing for me," and it moves me. But I can't think about that when I'm writing. I would be focused on the wrong place, the wrong side of the spectrum. I have a certain faith, I suppose, that if I trust my curiosity and I trust my passion for life, and only write about what fascinates me, then other people will share my curiosity. They will be able to see what I see and rejoice in it. I have that faith, just as I have a faith which is, I suppose, a kind of reverse fatalism, that when all is said and done, the work that I leave behind will have required the life that I led.

Rusty: Have you written for children?

Diane: I've written one book in the Crown Science Series for children. The illustrations are photographs. Many are priceless natural history photographs, and then there is text. I wrote one about monk seals. It's called *Monk Seal Hideaway*. I went to the South Pacific to write about monk seals for *National Geographic*. (It's part of a book that I'm doing called *The Rarest of the Rare*, a book about endangered animals.) I went out to see some of the most endangered animals in the world just to celebrate them and observe them and write about them before they disappear. It bothers me that we have the equivalent of dinosaurs walking around among us, and we fret over whether the ancient dinosaurs were cold-blooded or warm-blooded or what their skin was like or what they sounded like and did they run, did they trot. We don't know basic things about them, but we also have animals going extinct *now* that we don't know basic things about. Certainly, we have not appropriately mourned their passing, or taken responsibility for the fact that they're passing. I just feel better going out on behalf of everybody, and witnessing these animals before they disappear.

In the case of monk seals, I was very fortunate because I got to see monk seals mating under water. And this was only the third time in history it had been reported, and nobody had seen them in this kind of detail before. I looked down and there they were, doing what they do.

Rusty: When you were talking earlier about the books and the integration of parts that go into your books, I was also thinking that your prose books are full of information. In the children's book how did you figure out different levels of audience and how much information to give?

Diane: That's hard, and on occasion I got things wrong. When the edited manuscript came back to me, I was told, "Don't use the word 'altruism.' It's too sophisticated for these kids." But I was also told that, in the opening paragraph, I was talking too much to a juvenile level, and so I lifted the level of address there. This is quite new for me, and now I think I've reached the right balance and tone. If I do another children's book, and I probably will, I'll know a little bit better now.

Barbara: It seems to me that the "witnessing," as you called it when you were talking about observing animals, goes together with the bookish research, and that you always use them in tandem—but the witnessing has to be there.

Diane: I think that's so. I feel a combination of celebration, enquiry, witnessing, prayer, and all the seductive complexities of memory at the same time, and just rank curiosity, Faustian curiosity. All of this rolls together when I write something, but I never start a book because I know what I'm going to find out. There'd be no fun in that. Every book that I begin is a journey. It's an exploration into a subject that I find fascinating and that is a dark continent to me. It may open up and allow me to blaze trails here and there but will still retain mystery.

Rusty: Thinking as a historian about various nature traditions, I hear in you a kind of romantic tradition.

Barbara: And I hear Victorian women travelers. Have you read Mary Kingsley, or Isabella Bird's *A Lady's Life in the Rocky Mountains?*

Diane: And Lady Hester Stanhope, and later, Freya Stark. There was a good rough and ready, you know, a muck-in-and-let's-do-it attitude to women of that time. As for the romantic, there's a Whitmanesque attitude, and also Thoreau and Emerson. I feel spiritually close to Emerson's attitudes . . . not all of his attitudes about life, not all of his attitudes about women, not all of his attitudes about love. But the position he finds in nature is sometimes so discrete and so receptive that it speaks to me.

Rusty: I did wonder whether, in your own work, you saw differences between male science and female science?

Diane: I don't really know, and also the attitude that people have about male scientists and about female scientists gets in the way. When I published *The Moon by Whale Light* various women's magazines did profiles of me, and occasionally reporters would see me as a great risk taker, adventurer, doing heroic things. I had to remind them that there are field biologists, *women* field biologists out there every day doing similar things. For some reason, the minute people think of women out in the wild they see it as death-defying. They are not used to thinking of women in the same situations where they would calmly think of men.

But actually that's been going on all the time, and actually I don't take risks. I'm unbelievably careful when I set out on expeditions, and as far as I'm concerned, heroism is something that happens most often privately in people's homes, in families and in small daily acts. It doesn't necessarily have to do with deciding that you're going to go to the Amazon in a safe way, or to the Antarctic in a safe way, or something like that.

Barbara: Earlier, we were talking about women and the tradition of nature writing, and popularization—about women back in time.

Diane: Popularization . . . people want to understand the world around them, but they don't want another boring science lecture, they don't want to be talked down to, they don't want raw science offered in a way that's unrelated to their everyday lives, and they don't want jargon. They want a sense of the beauty of the world, a way to understand the mysteries that they move among, and they want respect while they're learning. So, it's a great time to be writing that beast that we don't have a name for—which some of us call the literary essay, and some of us call creative nonfiction writing—but which really has to do, in my case anyway, with popularizing science, or rather popularizing nature.

THE DARK NIGHT OF THE HUMMINGBIRD

A lot of hummingbirds die in their sleep,
dreaming of nectar-sweet funnels they sipped.
Moth-light, they swiveled at succulent
blooms, all flash and ripple—like sunset,
but delicate, probing, excitable,
their wings a soft fury of iridescence,
their hearts beating like a tiny drumroll
fourteen hundred times a minute,
their W-shaped tongues, drawing nectar
down each groove, whispering: *wheels within wheels*.
By day, hovering hard, they fly nowhere
at speed, swilling energy. But to refuel,
they must eat, and to eat they must hover,
burning more air than a sprinting impala.

So, in the dark night of the hummingbird,
while lilies lather sweetly in the rain,
the hummingbird rests near collapse,
its quick pulse halved, its rugged breath shallow,
its W-shaped tongue, bright as Cassiopeia,
now mumbling words like *wistful* and *wan*.

The world at once drug, anthem, bright lagoon,
where its heart knew all the Morse codes
for rapture, pales into a senseless twilight.
It can't store enough fuel to last the night
and hoist it from its well of dreams
to first light trembling on wet fuchsia,
nor break the hard promise life always keeps.
A lot of hummingbirds die in their sleep.

Diane Ackerman

Diane Ackerman (b. 1948): Major Works

Lady Faustus (1983)
A Natural History of the Senses (1990)
Jaguar of Sweet Laughter: New and Selected Poems (1991)
A Natural History of Love (1994)
The Rarest of the Rare: Vanishing Animals, Timeless Worlds (1995)

Selected Bibliography
Index

Selected Bibliography

Abir-Am, Pnina, and Dorinda Outram, eds. *Uneasy Careers and Intimate Lives: Women in Science 1789–1979*. New Brunswick: Rutgers UP, 1987.

Ainley, M. G., ed. *Despite the Odds: Essays on Canadian Women and Science*. Montreal: Véhicule P, 1990.

Alic, Margaret. *Hypatia's Heritage: A History of Women in Science from Antiquity through the Nineteenth Century*. Boston: Beacon P, 1986.

Allen, David Elliston. *The Naturalist in Britain*. 2d ed. Princeton: Princeton UP, 1995.

Ballaster, Ros, et al. *Women's Worlds: Ideology, Femininity, and the Woman's Magazine*. London: Macmillan, 1991.

Bazerman, Charles. *Shaping Written Knowledge: The Genre and Activity of the Experimental Article in Science*. Madison: U of Wisconsin P, 1988.

Bazerman, Charles, and James Paradis, eds. *Textual Dynamics of the Professions: Historical and Contemporary Studies of Writing in Professional Communities*. Madison: U of Wisconsin P, 1991.

Beer, Gillian. *Darwin's Plots: Evolutionary Narrative in Darwin, George Eliot, and Nineteenth-Century Fiction*. London: Routledge and Kegan Paul, 1983.

Benjamin, Marina. "Elbow Room: Women Writers on Science, 1790–1840," in Benjamin, *Science and Sensibility*.

Benjamin, Marina, ed. *A Question of Identity: Women, Science, and Literature*. New Brunswick: Rutgers UP, 1993.

Benjamin, Marina, ed. *Science and Sensibility: Gender and Scientific Enquiry, 1780–1945*. Oxford: Basil Blackwell, 1991.

Birke, Lynda, and Ruth Hubbard, eds. *Reinventing Biology: Respect for Life and the Creation of Knowledge*. Bloomington: Indiana UP, 1995.

Bonta, Marcia Myers. *Women in the Field: America's Pioneering Women Naturalists*. College Station: Texas A&M UP, 1991.

Brody, Judit. "The Pen Is Mightier Than the Test Tube." *New Scientist* 105 (1985): 56–58.

Cooter, Roger, and Stephen Pumfrey. "Separate Spheres and Public Places: Reflections on the History of Science Popularization and Science in Popular Culture." *History of Science* 32 (1994): 237–67.

Dale, Peter Allan. *In Pursuit of a Scientific Culture: Science, Art, and Society in the Victorian Age.* Madison: U of Wisconsin P, 1989.

Douglas, Aileen. "Popular Science and the Representation of Women: Fontenelle and After." *Eighteenth-Century Life* 18 (1994): 1–14.

Ezell, Margaret J. M. *Writing Women's Literary History.* Baltimore: Johns Hopkins UP, 1993.

Gaard, Greta, ed. *Ecofeminism: Women, Animals, Nature.* Philadelphia: Temple UP, 1993.

Gates, Barbara T. "Retelling the Story of Science." *Victorian Literature and Culture* 21 (1993): 289–306.

Haraway, Donna J. "A Game of Cat's Cradle: Science Studies, Feminist Theory, Cultural Studies." *Configurations: A Journal of Literature, Science, and Technology* 1 (1994): 59–71.

Haraway, Donna J. *Simians, Cyborgs, and Women: The Reinvention of Nature.* New York: Routledge, 1991.

Harding, Sandra. *Whose Science? Whose Knowledge? Thinking from Women's Lives.* Ithaca: Cornell UP, 1991.

Harth, Erica. *Cartesian Women: Versions and Subversions of Rational Discourse in the Old Regime.* Ithaca: Cornell UP, 1992.

Hobby, Elaine. *Virute of Necessity: English Women's Writing, 1649–88.* Ann Arbor: U of Michigan P, 1989.

Jordanova, Ludmilla. "Gender and the Historiography of Science." *British Journal for the History of Science.* 26 (1993): 469–83.

Jordanova, L. J. *Languages of Nature: Critical Essays on Science and Literature.* New Brunswick: Rutgers UP, 1986.

Jordanova, Ludmilla. *Sexual Visions: Images of Gender in Science and Medicine between the Eighteenth and Twentieth Centuries.* Madison: U of Wisconsin P, 1989.

Keeney, Elizabeth B. *The Botanizers: Amateur Scientists in Nineteenth-Century America.* Chapel Hill: U of North Carolina P, 1992.

Keller, Evelyn Fox. *Reflections on Gender and Science.* New Haven: Yale UP, 1985.

Keller, Evelyn Fox. *Secrets of Life, Secrets of Death: Essays on Language, Gender, and Science.* New York: Routledge, 1992.

Koerner, Lisbet. "Women and Utility in Enlightenment Science." *Configurations* 2 (1995): 233–55.

Kohlstedt, Sally Gregory. "In from the Periphery: American Women in Science, 1830–1880" *Signs* 4 (1978): 81–96.

Latour, Bruno. *We Have Never Been Modern.* Trans. Catherine Porter. Cambridge: Harvard UP, 1993.

Levine, George, ed. *Realism and Representation: Essays on the Problem of Realism in Relation to Science, Literature, and Culture.* Madison: U of Wisconsin P, 1993.

Locke, David. *Science as Writing.* New Haven: Yale UP, 1992.

Longino, Helen. *Science as Social Knowledge: Values and Objectivity in Scientific Enquiry.* Princeton: Princeton UP, 1990.

McRae, Murdo William, ed. *The Literature of Science: Perspectives on Popular Scientific Writing.* Athens: U of Georgia P, 1993.

Merchant, Carolyn. *The Death of Nature: Women, Ecology, and the Scientific Revolution*. San Francisco: Harper & Row, 1980.

Merchant, Carolyn. *Earthcare: Women and the Environment*. New York: Routledge, 1996.

Meyer, Gerald Dennis. *The Scientific Lady in England, 1650–1760: An Account of Her Rise, with Emphais on the Major Roles of the Telescope and Microscope*. Berkeley: U of California P, 1955.

Morgan, Susan. *Place Matters: Gendered Geography in Victorian Women's Travel Books about Southeast Asia*. New Brunswick: Rutgers UP, 1996.

Moscucci, Ornella. *The Science of Woman: Gynecology and Gender in England, 1800–1929*. New York: Cambridge UP, 1990.

Mullen, John. "Gendered Knowledge, Gendered Minds: Women and Newtonianism, 1690–1760," in Benjamin, *A Question of Identity*.

Myers, Greg. "Fictions for Facts: The Form and Authority of the Scientific Dialogue." *History of Science* 30 (1992): 221–47.

Myers, Greg. "Science for Women and Children: The Dialogue of Popular Science in the Nineteenth Century," in *Shuttleworth and Christie*.

Myers, Greg. *Writing Biology: Texts in the Social Contruction of Scientific Knowledge*. Madison: U of Wisconsin P, 1990.

Myers, Mitzi. "Aufklarung für Kinder? Maria Edgeworth and the Genders of Knowledge Genres; Or, 'The Genius of Nonsense' and 'The Grand Panjandrum Himself'," *Women's Writing: The Elizabethan to Victorian Period* 2, no. 2 (1995): 113–40 (Special Issue "Women and Science").

Myers, Mitzi. "Impeccable Governesses, Rational Dames, and Moral Mothers: Mary Wollstonecraft and the Female Tradition in Georgian Children's Books." *Children's Literature* 14 (1986): 31–59.

Norwood, Vera. *Made from This Earth: American Women and Nature*. Chapel Hill: U of North Carolina P, 1993.

Phillips, Patricia. *The Scientific Lady: A Social History of Woman's Scientific Interests, 1520–1918*. London: Weidenfeld & Nicolson, 1990.

Pycior, Helena M., Nancy G. Slack, and Pnina G. Abir-Am, eds. *Creative Couples in the Sciences*. New Brunswick: Rutgers UP, 1996.

Rossiter, Margaret. *Women Scientists in America: Before Affirmative Action, 1940–1972*. Baltimore: John Hopkins UP, 1995.

Rossiter, Margaret. *Women Scientists in America: Struggles and Strategies to 1940*. Baltimore: Johns Hopkins UP, 1982.

Russett, Cynthia Eagle. *Sexual Science: The Victorian Construction of Womanhood*. Cambridge: Harvard UP, 1989.

Schiebinger, Londa. *The Mind Has No Sex? Women in the Origins of Modern Science*. Cambridge: Harvard UP, 1989.

Schiebinger, Londa. *Nature's Body: Gender and the Making of Modern Science*. Boston: Beacon P, 1993.

Shevelow, Kathryn. *Women and Print Culture: The Construction of Femininity in the Early Periodical*. London: Routledge, 1989.

Shinn, Terry, and R. Whitley. *Expository Science: Forms and Functions of Popularisation*. Dordrecht: D. Reidel, 1985.

Shteir, Ann B. *Cultivating Women, Cultivating Science: Flora's Daughters and Botany in England, 1760 to 1860*. Baltimore: Johns Hopkins UP, 1996.
Shteir, Ann B., ed. Special Issue "Women and Science." *Women's Writing: The Elizabethan to Victorian Period*. 2, no. 2 (1995).
Shuttleworth, Sally, and J. R. R. Christie, eds. *Nature Transfigured: Literature and Science, 1700–1900*. Manchester: Manchester UP, 1989.
Tuana, Nancy, ed. *Feminism and Science*. Bloomington: Indiana UP, 1989.

Index

Abir-Am, Pnina, 4
Aborigines, 104, 106, 111–12, 114*n3*
Ackerman, Diane, 21, 255–64
Adams, J. F. A., 17
Adaptation, Lamarckian conception of, 150–51
Aesthetics, 35, 119, 128, 130–34, 138. *See also* Beauty
Africa, 215–30, 237–51
African Madness (Shoumatoff), 242
Agassiz, Louis, 126
Agriculture, 82, 120–21, 138, 156, 199. *See also* Pesticides
AIDS, 238
Aikin, Joan, 47
Ainley, Marianne Gosztonyi, 4, 19, 79–97
Alaya, Flavia, 159
Alexander, John, 244
Algebra, 6, 72
Allen, D. E., 28
Altruism, 20, 152, 159, 165, 261. *See also* Sympathy
Always Coming Home (Le Guin), 47
Amazon revolt, 157
American Association for the Advancement of Science, 222
Androcentrism, 79–80
Anglo-American Magazine, 86
Animism, 66
Anning, Mary, 29
Anonymity, of the reviewer, 51–52
Anthropology, 15, 106–8; and evolution, 148–49, 154–56, 160; and Kingsley, 217, 221–23
Anthropomorphism, 122, 125
Apes. *See* Chimpanzees; Gorillas; Orangutans
Aquarium (Gosse), 10–11
Arab world, 5
Aristocracy, 48
Armstrong, Nancy, 159
Astronomy, 5–8, 19, 61–75, 174–75
At Last: A Christmas in the West Indies (Kingsley), 108
Athenaeum, 36, 102
Athenian Mercury, 5
Atomic Energy Commission, 209
Australia, 4, 19, 98–115

Bachofen, Johann Jakob, 157
Backwoods of Canada (Traill), 82–86, 105
Bailey, Florence Merriam, 93
Bailey, Liberty Hyde, 126
Baker, W. C., 130
Bakerian Lectures, 48–52
Ball, Robert S., 71
Ballstadt, Carl A., 80
Barber, Lynn, 29–30
Barrias, Louis Ernest, 9
Beauty, 32–36, 204, 263; perception of, in Davy, 50; and evolution, 151–52. *See also* Aesthetics
Beer, Gillian, 161*n2*, 170
Behn, Aphra, 5
Bell, Thomas, 101–2, 105, 108
Benevolence, 164. *See also* Altruism; Sympathy
Benjamin, Marina, 4
Berger, Carl, 80
Bible, 38, 202. *See also* Christianity; God

Bird, Isabella, 262
Bird behavior, 84–85
Birds of Europe (Gould), 29
Birth control, 20, 180, 184
Black Hearts at Battersea, 47
Blackwell, Antoinette Brown, 150
Bleak House (Dickens), 47
Bodies, women's, 7, 151. *See also* Sexuality
Bodington, Alice, 14–15
Botanical Society of London, 28–29
Botany, 7–8, 10, 13–14; and Traill, 19, 79–97; and Meredith, 98–115
Botsford, Phoebe, 116
Bowler, Peter, 134
Boyle, Kay, 50, 192
Boy's Own Book of Natural History (Wood), 16
Brightwen, Eliza, 16
Brindamour, Rod, 249
Britain, 4, 10, 13–14, 43–75
British Association for the Advancement of Science, 64
British Astronomical Association, 74*n2*
British Seaweeds (Gatty), 14
British Wild Flowers (Loudon), 13
Brockway, Lucile H., 100–101, 108
Brooks, Paul, 198, 205
Brooks, W. K., 150
Brougham, Henry, 51–53, 57
Bryan, Margaret, 8, 63
Bryce Commission, 17
Buckley, Arabella, 15–16, 20, 164–76
Buhl, Mari Jo, 161*n10*
Bureau of Fisheries, 200
Burke, Edmund, 35–36, 39*n1*
Burton, Richard, 219
Butterfly Hunters, The (Treat), 118
Byron, Lord, 106, 119, 128, 130

Caitling, P. M., 93
Campanella, Tommaso, 66
Campbell, Bob, 243, 244
Canada, 19, 79–97
Canadian Crusoes (Traill), 83, 93
Canadian Wild Flowers (Traill), 85, 87–90, 93
Cannibalism, 226
Cape of Good Hope observatory, 70
Capitalism, 158, 161*n10*, 237
Carson, Rachel, 18, 20–21, 196–211
Catalogue of Canadian Plants (Macoun), 90
Catholicism, 63, 66. *See also* Religion
Cavendish, Margaret, 6, 19
Central Experimental Farm, 90
CERN, 51
Chambers, Robert, 102
Chautauqua movement, 127
Chautauquan, 127
Chemistry, 7–8, 18–19, 43–60, 66, 85
Chernobyl nuclear accident, 208
Children's education, 7–9, 12, 31; and Clerke, 63; and Comstock, 116–19, 121–24, 126–27, 136, 138–39. *See also* Education
Chimpanzees, 237–39
Chorley, Henry, 104
Christianity, 63, 66, 157, 224–25. *See also* God; Religion
Church Quarterly, 228
Clark, Suzanne, 192
Class, 46–48, 82–83, 106, 158; middle class, 46, 239; working class, 46, 47, 218
Clerke, Agnes Mary, 19, 44, 61–75
Clerke, Ellen Mary, 64
Clifford, James, 161*n2*
Cole, Jean M., 79
Coleridge, Samuel Taylor, 173
Colette, Sidonie-Gabrielle, 255
Collins, Jane L., 243, 248, 250
Cologne Zoo, 250
Colonialism, 81, 109, 111–13, 237, 243
Comments on Birth Control (Mitchison), 180, 184
Committee for the Promotion of Agriculture, 126
Communist Party, 27
Compendious System of Astronomy, A (Bryan), 8, 63
Comstock, Anna Botsford, 19, 116–43
Comstock, John Henry, 19, 120–25, 134–35, 137, 139
Comstock Publishing Associates, 125
Comstocks of Cornell, The (Comstock), 119
Conchologist's Companion (Roberts), 18, 30–39
Conchology, 18, 30–39
Confessions to a Heathen Idol (Comstock), 121

Conrad, Joseph, 241–42
Conversations on Chemistry (Marcet), 8, 18–19, 45–60
Conversations on Natural Philosophy (Marcet), 8, 45
"Cool Debate, A" (Meredith), 108–11
Cooter, Roger, 3–4
Copernicus, Nicolaus, 65
Cornell Nature-Study Leaflets, 127
Cornell University, 119, 121, 125, 126
Corson, Hiram, 119
Country Life in America, 127
Crime, 223–34
Critic, 159
Croft, L. R., 82
Crown Science Series, 260
Cullis, Winifred, 189
Cuvier, Georges, 38

Daedalus, or Science and the Future (Haldane), 182–83, 185
"Dark Night of the Hummingbird" (Ackerman), 258–59, 263–64
Dark Romance of Dian Fossey, The (Hayes), 244
Darlington, William, 82
Darwin, Charles, 5, 11–12, 15, 87, 134; and Gamble, 20, 148–55, 158–59; Buckley's critique of, 20, 164–76; Gould on, 32; and Clerke, 67, 73; and Bell, 101; and Meredith, 105, 107; *The Descent of Man, and Selection in Relation to Sex*, 107, 137, 148–51, 153, 158, 169; and Kingsley, 221, 232*n14*
Darwin, Erasmus, 164
Davis, John P., 120
Davy, Humphry, 18, 45–57
Dawson, J. W., 87
Day, Thomas, 81
DDT, 204–6. *See also* Pesticides
Defoe, Daniel, 81
Demonstration, 43–60
Department of Agriculture, 120
Descent of Man, and Selection in Relation to Sex, The (Darwin), 107, 137, 148–51, 153, 158, 169
Dial, 228
Dickens, Charles, 12, 47, 119
Digit Fund, 241
Dillard, Annie, 200–201
Discourse of Difference (Mills), 105
Discovery of New Worlds (Behn), 5
Domestic Recreation; or, Dialogues Illustrative of Natural and Scientific Objects (Wakefield), 11
Don Juan (Byron), 106
Donne, John, 255
d'Orsay, Count, 106

Earl of Minto, 48
Early, Julie English, 21, 215–36
Easy Introduction to the Knowledge of Nature, and Reading the Holy Scriptures, adapted to the capacities of children (Trimmer), 7–8
Ecology, 85, 88, 93. *See also* Environmentalism
Edinburgh Review, 50–51, 53, 57, 65, 217, 224
Edinburgh University, 81
Education, 7–8, 10–11, 137–39; introduction of natural history, in schools, 17; and Comstock, 125–26, 199. *See also* Children's education
Egalitarianism, 159
Electricity, 5
Eliot, George, 10
Engels, Friedrich, 155–56, 161*n10*
"Englishness, Travel and Theory: Writing the West Indies in the Nineteenth Century" (Gikandi), 100
Enlightenment, 5–7, 159
Entomology, 6–8, 10, 19–20, 116, 166
Entretiens sur la pluralité des mondes (Fontenelle), 5
Environmentalism, 18, 20–21, 196–211. *See also* Ecology
Epistemology, 20, 189
Essay on Man (Pope), 34–35
Essentialism, 149, 151, 181, 193*n2*
Eternal rhythms, 202–3
Ethnology, 107, 137, 153–54, 221, 226–27
Evolution, 87, 137, 147, 148–63, 165–75; and the survival of the fittest, 134–36; and anthropology, 148–49, 154–56, 160
Evolution of Sex, The (Geddes and Thompson), 165
Evolution of Woman, The (Gamble), 149, 158

Fairy-Land of Science, The (Buckley), 170–74
Faraday, Michael, 46
Fee, Elizabeth, 153
Female Emigrant's Guide, The (Traill), 85
Female Spectator, The, 6–7
Femininity, 4, 7, 43; and Roberts, 34, 35; and Comstock, 119; and evolution, 147–63. *See also* Gender
Feminism, 20, 27–28, 179–95
Fenn, Lady Elleanor, 118
Ferguson, Kathy, 193
Fetish, 222
Fictionality, 43–60
FitzGibbon, Agnes Moodie, 88, 90
Fletcher, James, 90, 93
Flora Americae Septentrionalis (Pursh), 84, 86, 92
Folk-Lore, 224, 228
Forbes, O. L., 130
Ford, Laetitia, 118
Fossey, Dian, 21, 237–51
Franklin, Christine Ladd, 159
Frazer, James George, 192, 224, 232*n16*
Freud, Sigmund, 160
Fuertes, Louis Agassiz, 130

Gage, Simon H., 120
Gage, Susanna Phelps, 120
Gaia hypothesis, 169–70
Galdikas, Biruté, 21, 237–51
Galileo, 54
Gamble, Eliza Burt, 15, 20, 147–63
Gamble, James, 149
Gates, Barbara T., 3–24, 164–76, 255–64
Gatty, Margaret, 14
Geddes, Patrick, 151, 165
Gender, 4, 7–10, 21; and scientific research, Gould on, 27–39; and Marcet, 43, 46–60; and Clerke, 71–73; and Comstock, 139–40; and evolution, 147–63; and scientific feminisms, 182–93; and primatology, 237, 238, 246–50
General Magazine of Arts and Sciences, The, 6
Genesis, 202
Gentleman's Magazine, 5
Geological and Natural History Survey of Canada, 90
Geological Society of London, 28
Geology, 10, 12–14, 28, 43, 90; and Roberts, 38; and Traill, 82, 84, 87; and Buckley, 165
Gikandi, Simon, 100, 103, 108
Gill, David, 70
Gilman, Charlotte Perkins, 158
"Gland That Controls Your Sex, The" (Holmes), 186
Glaucus; or, The Wonders of the Shore (Kingsley), 10, 11, 16
God, 32–34, 37, 89, 149, 157; truth of and the truth of science, 14; and astronomy, 63, 67–69; and Traill, 89, 91; evidence of, in nature, 168–69. *See also* Natural theology; Religion
God-Idea of the Ancients, The (Gamble), 149, 157
Golden Bough, The (Frazer), 192, 224
Golinski, Jan, 47–48
Goodall, Jane, 237–39
Goodness, universal, 32–33
Gorillas, 237–45, 250–51
Gorillas in the Mist (Fossey), 239, 241, 243
Gosse, Edmund, 16–17
Gosse, Philip, 10–11
Gould, John, 29
Gould, Stephen Jay, 18, 27–39
Graham, Frank, 204
Gravity, 5, 66, 170
Gray, Asa, 87, 89
Great Exhibition of London, 111
Green, Alice, 220, 230*n2*
Gregory, Leonora, 186–87
Gregory, Richard A., 72, 73
Grey, Maria, 164
Griffiths, Mrs. A. W., 29
Grounds of Natural Philosophy (Cavendish), 6
Guide to the Study of Insects (Packard), 124
Guillemard, Henry, 218, 227, 233*n20*
Gulf Oil, 237

Haeckel, Ernst, 66
Hage, Ghassan, 110–11
Haggard, H. Rider, 241
Haldane, Charlotte, 20, 179–95
Haldane, J. B. S., 20, 179, 182, 185–86
Haldane, John Scott, 179, 183
Hall, Sarah Aiken, 118
Hammond, Dorothy, 240

Handbook of Nature Study (Comstock), 126–27, 138
Haraway, Donna, 160, 237, 240, 244–45, 249
Harper's Weekly, 118
Hayes, T. P., 244
Haywood, Eliza, 6
Henson, Pamela M., 116–43
Herbicides, 196. *See also* Pesticides
Herschel, Caroline, 64, 65, 73
Hesse-Honegger, Cornelia, 207–8, 210
Heterosexuality, 7, 192. *See also* Sexuality
Hincks, William, 87, 89
Historical and Miscellaneous Questions for the Use of Young People (Mangnall), 9
Hoberman, Ruth, 192
Hogg's Weekly Instructor, 103
Holden, E. S., 70
Holmes, Barbara, 186
Home Studies in Nature (Treat), 118
Hooker, W. J., 82, 88, 92
Hopkins, John Gowland, 186
Horticulturalist (Traill), 86, 92
Housekeeping (Robinson), 208
"How to Write a Popular Scientific Article" (Haldane), 182
Huggins, Margaret, 61–64, 69–71, 73
Huggins, William, 61
Huxley, Thomas Henry, 11, 165, 169

Illustrations, 13, 19–20, 90, 118–23, 128–34
Illustrations of the Natural Orders of Plants with Groups and Descriptions (Twining), 13
Imagination, 61, 173–74
Imperialism, 5, 229
Indian Mutiny, 111
Indigenous peoples, 104, 106, 111–12, 114*n3*, 221–22, 226
Infanticide, 153, 156–57
In Memoriam (Tennyson), 102
Insect Life (Comstock), 122, 125, 130, 132–33
Insects. *See* Entomology; Pesticides
Instinct, 152, 155, 159, 169, 221, 227
Intelligence, 15, 27–28, 151–52
Inter-Colonial Exhibition, 111
International Wildlife, 246
Introduction to Botany, An (Wakefield), 8
Introduction to Entomology (Comstock), 121
Introduction to the Natural History and Classification of Insects, An (Wakefield), 118
"Is Botany a Suitable Study for Young Men?" (Adams), 17

Jablow, Alta, 240
Jack, Annie L., 85
Jacson, Maria, 46, 118
James the Third (king), 47
Jann, Rosemary, 20, 147–63
Jérémie, Catherine, 82
Jevons, F. B., 224, 225
Johns Hopkins University, 150, 199
Johnston, Judith, 19, 98–115
Jones, Eliza M., 85
Jordanova, Ludmilla, 9
Joyce, Jeremiah, 46
Jussieu, Antoine Laurent de, 89

Keats, John, 128, 130
Keller, Evelyn Fox, 119, 130
Kelvin, Lord, 69
Kemp, Dennis, 220
Kessler, Karl, 169
Khrushchev, Nikita, 27
Kingsley, Charles, 10, 16–17, 29, 108
Kingsley, George, 218
Kingsley, Mary, 12, 21, 215–36, 262
Kirby, William, 87
Knight, David, 46
Kohlstedt, Sally Gregory, 118, 136
Krasner, James, 21, 170, 172, 237–51
Kropotkin, Peter, 169

Ladies' Diary, 6
Ladies' Flower Garden of Ornamental Annuals (Loudon), 13
Lady Mary and Her Nurse (Traill), 83
"Lady of Shalott, The" (Tennyson), 105
Lady's Life in the Rocky Mountains, A (Bird), 262
Lady's Museum, The, 6
"Lady Travellers" (Rigby), 102–4, 113
Lamarck, Jean-Baptiste, 150–51
Last Series: Bush Friends in Tasmania, Native Flowers, Fruits and Insects, Drawn from Nature with Prose Descriptions and Illustrations in Verse (Meredith), 108–12

Latham, Robert, 107
Latin names, 85, 136
Lawrence, Thomas, 98
Lawson, George, 87, 89
League of Women Voters, 127
Leakey, Louis, 237, 239, 250
Lear, Edward, 29
Lectures on Natural Philosophy (Bryan), 8
Le Guin, Ursula, 47
Lennox, Charlotte, 6
Leopold, Aldo, 196
Levander, F. W., 74*n2*
Lévi-Strauss, Claude, 160
Lewes, George Henry, 10
Lewis, Graceanna, 93
Lewis, Jane, 182
Life and Her Children (Buckley), 165–66, 169, 170, 172
Life and Thoughts of a Naturalist (Brightwen), 16
Lightman, Bernard, 19, 61–75
"Lilies" (Traill), 92
Linnaean system of classification, 13, 102, 107, 136
Linnaeus, Carl, 89
"Linnaeus's Daughters: Women and British Botany" (Shteir), 101
Linnean Society, 28
"Lipsticks are Politics" (Gregory), 186–87
Liverpool Geographical Society, 227
Lockyer, Norman, 69, 72
London Quarterly Review, 220
Loudon, Jane, 12–13
Loudon, John Claudius, 13
Love, 121, 159, 184, 192
Lovelock, James, 170
Lubbock, John, 153, 155–57
Lutz, Catherine A., 243, 248, 250
Lyall, Alfred, 224–25
Lyell, Charles, 12, 43, 102, 165, 172, 174–75
Lynam's School, 183
Lyon, Thomas, 198

McCallum, Elizabeth, 79
McCloskey, Alice, 126
Macdonald, Lady, 218–19
McLennan, John Ferguson, 153, 155–56
Macmillan Publishers, 215, 217–18, 220, 224–26, 227
Macoun, John, 87, 90
Madame How and Lady Why (Kingsley), 12
Magazine of Natural History, 12
Magnetism, 5, 170
"Making Friends with the Mountain Gorilla" (Fossey), 239–40
"Mammalia: Extinct Species and Surviving Forms, The" (Bodington), 15
Mangnall, Richmal, 9
Mann, R. J., 71
Manual for the Study of Insects (Comstock), 124, 129–31
Marcet, Jane, 8, 18–19, 43–60
Margulis, Lynn, 170
Marie de l'Incarnation, 82
Marriage, 121, 154–56, 158
Martin, Benjamin, 6
Martineau, Harriet, 18, 57
Masculinity, 4, 17, 43; and Roberts, 34, 35, 37; and Comstock, 119; and evolution, 150–60. *See also* Gender
Mating behavior, 136–37, 249
Matriarchy, 153–54, 157–59
Maunder, E. W., 71
Meadows, A. J., 74*n2*
Melbourne Exhibition, 112
Meliorism, scientific, 181, 187, 192
Mental Improvement, or the Beauties and Wonders of Nature and Art (Wakefield), 8
Merchant, Carolyn, 9
Meredith, Louisa Anne (Louisa Anne Twamley), 19, 98-115
Middle class, 46, 239
Millay, Edna St. Vincent, 192
Mills, Sara, 105
Mitchell, Mary, 70
Mitchison, Naomi, 20, 179–95
Modern Cosmogonies (Clerke), 68–70
Modernism, 191–92
Monism, 66
Monk Seal Hideaway (Ackerman), 261
Monogamy, 156
Montgomery, Sy, 243, 245, 247, 249–50
Monthly Review, 101
Montreal Daily News, 90
Moon by Whale Light, The (Ackerman), 262
Morality, 8, 11, 164; and Comstock, 118, 137–38; and evolution, 152, 153; and Buckley, 164–65, 168–69

Moral Teachings of Science (Buckley) 168–69
"More Years with the Mountain Gorilla" (Fossey), 240, 244
Morgan, Lewis Henry, 155, 156, 161*n10*
Morris, Maria, 87, 88
Mother Country (Robinson), 208
Motherhood, 7, 122, 152, 246–48; metaphors for, 86; and Haldane, 180–84, 190
Motherhood and Its Enemies (Haldane), 180, 184, 190
Mowat, Farley, 250
Muir, John, 196
Müller, F. Max, 224
Murchison, Roderick, 217
Murray, John, 102
"Mute Dancers: How to Watch a Hummingbird" (Ackerman), 258, 263–64
Mutual aid, 20, 169
Mutual Aid: A Factor in Evolution (Kropotkin), 169
Myers, Greg, 11, 18–19, 43–60
Myers, Mitzi, 9

Narrative form, 9, 11–13, 147, 245–46
Nation, 71, 159, 231*n7*
National Book Award, 198
National Geographic, 21, 237–39, 243–46, 248, 250, 261
Natural History of Love, A (Ackerman), 260
Natural History of the Senses, A (Ackerman), 257, 260
Natural theology, 11–13, 15–18, 67–68, 73
Natural Theology (Paley), 11–12
Nature, 51, 65, 73, 228
Nature Magazine, 128
"Nature Musings from a Car Window" (Comstock), 128
Nature-Study Idea (Bailey), 126
Nature study movement, 126
"Nature unveiling herself before Science" (sculpture), 9
Needler, G. H., 79
Nevada Test Site, 209–10
New Century, 90
New College, 183
New Science, 259
Newton, Isaac, 5, 38, 66, 175, 257
New York Review of Books, 50
New York State Geological Survey, 118
New York Times, 258, 259
Nineteenth Century, 169
Nobel Prize, 259
Notes and Sketches of New South Wales (Meredith), 100, 102–4, 106, 109–10
Nova Scotia Wildflowers (Morris), 88
Nuclear power plants, 207–8
Nworah, Kenneth Dike, 233*n23*

Objectivity, 130–31, 217
Observations upon Experimental Philosophy (Cavendish), 6
"Old Story of 1834, An" (Meredith), 112
On the Connexion of the Physical Sciences (Somerville), 63–64
"On the Law of Mutual Aid" (Kessler), 169
Orangutans, 237, 239, 245–51
Original sin, 246
Origin myths, 147
Origin of the Family, Private Property, and the State (Engels), 156
Ornithology, 84–85
Our Mutual Friend (Dickens), 12
Outline for Boys and Girls and Their Parents, An, 184, 187, 188, 190–92
Outlines of Astronomy (Herschel), 65
Outram, Dorinda, 4
Ownership, of property, 155–56
Oxford University, 183
Oxygen, 52, 53, 56–57
Oxymuriatic acid, 48

Packard, Alpheus Spring, 124, 130
Paleontology, 29, 37–38
Paley, William, 11–12, 168
Parables of Nature (Gatty), 14
Paradigms, 11
Park, Mungo, 219
Patriarchy, 149, 152–54, 187
Pavel, Thomas, 47
Pennsylvania College for Women, 199
Penny Cyclopaedia, 101, 102, 104
Perceval, Anne Mary, 82
Perry, Ruth, 7
Pestalozzian method, 125
Pesticides, 196, 203–7
Peterman, Michael A., 80
Phelps, Almira H. L., 93

Phillips, Patricia, 14, 17
Philosophical Enquiry into the Origin of Our Ideas of the Sublime and Beautiful (Burke), 35–36
Philosophical Transactions, 48
"Philosophy for Ladies" series, 6–7
Physics, 66–67
Pilgrim at Tinker Creek (Dillard), 200
Pirie, N. W., 190
Plan for the Conduct of Female Education in Boarding Schools, 164
Poetry, 19, 21, 119, 127–28; and Meredith, 110–13; and Carson, 198; and Ackerman, 255, 257–58, 263–64
Pope, Alexander, 34–35
Popular History of Astronomy during the Nineteenth Century (Clerke), 65, 67–68, 70–71
Popular Science Monthly, 158–59
Porter, Dennis, 107–8
Postmodernism, 47, 58, 182, 193*n2*
Potash, 49–50, 52, 54, 56
Potassium, 45, 49–50, 53, 55–56
Potter, Beatrix, 18
Pratt, Anne, 10
Primatologists, 21, 160, 237–51
Principles of Geology (Lyell), 43, 102
Problems in Astrophysics (Clerke), 65, 69, 70-72
Progress of Creation (Roberts), 37–38
Progressive principle, 156
Promiscuity, 153, 154–55
Property, private/communal, 155–56
Pumfrey, Stephen, 3–4
Punch, 227
Pursh, Frederick, 84, 86, 88

Quarterly Review, 102–3
Quatermain, Allan, 241

Race, 107, 158, 237–38, 249–50. *See also* Racism
Racism, 5, 111, 113, 222, 232*n14*. *See also* Race
Radiation, 207–10
Raglon, Rebecca, 20, 196–211
Rape, 248–49
Rarest of the Rare, The (Ackerman), 261
RAS Council, 64
Reading National Geographic (Collins and Lutz), 250
Realist: A Journal of Scientific Humanism, 188
Recreations in Physical Geography, or the Earth As It Is (Zornlin), 13–14
Redmond, Ian, 244–45
Refuge (Williams), 209
Religion, 38, 157, 202–3, 224–25; and astronomy, 63, 66, 67, 69, 73; and evolution, 147–49. *See also* Christianity; God; Natural theology
Renaissance, 66
Rhythms, eternal, 202–3
Riding, Laura, 192
Rigby, Elizabeth, 102–4, 113
Roberts, Mary, 18, 21, 30–39
Robinson, Marilynne, 208–10
Romance of Nature, or, the Flower Seasons Illustrated (Meredith), 98–100
Rose, Flora, 126
Royal Astronomical Society, 64
Royal Geographical Society, 215–17, 219, 228
Royal Institution, 19, 47–48, 53, 57
Royal Society, 48, 53, 102
Russett, Cynthia, 149
Russia, 27
Rwanda, 237–51
Ryan, Marie-Laure, 47

Sagan, Carl, 258
St. Nicholas, 118, 121, 199
Say, Lucy, 118
Schiebinger, Londa, 182
Science, 17
Science and Colonial Expansion (Brockway), 100–101
Science News, 259
Scientific Lady, The (Phillips), 14, 17
Scientific Mother, 9–10
Scientism, 5
Scotland, 86
Scottish Geographical Magazine, 227
Sea Around Us, The (Carson), 198, 201–3
Seaside Studies (Lewes), 10
Secularization, 66
Sellafield nuclear waste plant, 208, 209
Semiology, 250
Semiotics, 238

Senate, 199
Sense of Wonder, A (Carson), 206
Seton, Ernest Thompson, 93
Sexes in Science and History, The (Gamble), 20, 149–50
Sexes throughout Nature, The (Blackwell), 150
Sexism, 35
Sexuality, 7, 20, 221–22; and mating behavior, 136-37; and evolution, 147, 149–50, 154–58, 160, 221–22; and scientific feminisms, 180, 184, 190, 192; and primatology, 238, 244, 246–49
Shapin, Steven, 50
"Sheep in Wolf's Clothing" (Comstock), 134
Sheepshanks, Anne, 64
Shelley, Mary, 18
Sheppard, Harriet Campbell, 82
Shirreff, Emily, 164
Short History of Natural Science (Buckley), 165
Shoumatoff, Alex, 242, 244
Showalter, Elaine, 9
Shteir, Ann B., 3–24, 102, 136, 255–64; on women as cultivators of science, 101; on the preparation of illustrations by women, 118
Silent Spring (Carson), 20–21, 196, 198–99, 204–8
Simond, Louis, 48
Sin, 246
Skepper, Charles, 187
Skinker, Mary Scott, 199
Slavery, 157–58
Smith, Titus, 87
Socialism, 27, 182, 185
Society for the Diffusion of Useful Knowledge, 45, 50, 101
Solution Three (Mitchison), 183
Somerville, Mary, 18, 45, 57, 63–64
Soul, 34
Spectator, 226, 227
Spencer, Herbert, 150, 155
Squier, Susan, 20, 179–95
Stanhope, Lady Hester, 262
Stark, Freya, 262
Stereotypes, 30, 192
Stewart, Frances, 84
Stewart, Kelly, 242
Stocking, George, 106–7, 111, 153
Story of the Heavens, The (Ball), 71
"Story We Love Best, The" (Comstock), 122
Strachey, J. St. Loe, 227
Strauss, Eric, 189
Studies in Evolution and Biology (Bodington), 14–15
Studies of Plant Life in Canada (Traill), 85, 87, 90–93
Submariners, 186
Survival of the fittest, 134–36
Sweden, 33
Sympathy, 164–65, 167, 169, 246, 248. *See also* Altruism
System of the Stars, The (Clerke), 65, 66, 70, 72

Tait's Edinburgh Magazine, 102
Tale of Mrs. Tiggywinkle (Potter), 47
Tanka the Otter (Williamson), 200
Tasmania, 101, 108–14
Tasmanian Journal of Natural Science, 101, 113
Taunton Commission, 17
Taxonomy, 85–86, 88, 92–93, 134
Teleology, 32–33, 136, 147–48
Tennyson, Alfred Lord, 102, 105, 119
Theology, natural, 11–13, 15–18, 67–68, 73
Thetis experiment, 186
"They Were Two Hours from Death but I Was Not Afraid" (Haldane), 186, 188
Things of the Sea Coast (Pratt), 10
Thomas, Lewis, 255
Thomson, J. Arthur, 151, 165
Thoreau, Henry David, 196, 262
Thoughts on Self-Culture, Addressed to Women (Grey and Shirreff), 164
Through the Magic Glasses (Buckley), 170, 174-75
Torgovnick, Marianna, 238
Torrey, John, 82
Traill, Catherine Parr, 19, 79–97, 105
Traill, Thomas, 81, 83, 87
Transactions, 82
Travels in West Africa (Kingsley), 215, 217-24, 227–29
Travel writing, 98--115, 127–28, 215–36, 262

Treat, Mary, 118
Trimmer, Sarah, 7–8
Trinomial nomenclature, 92
Trojan Ending (Riding), 192
Trojan War, 192
Truth, 14, 72, 112, 223; and evolutionary theory, 137, 159; Ackerman on, 259
Twamley, Louisa Anne (Louisa Anne Meredith), 19, 98-115
Twining, Elizabeth, 13
Tylor, E. B., 220–22, 224, 228, 230*n2*, 232*n15*
Tylor, Mrs. E. B., 220–22, 224, 228

Under the Sea Wind (Carson), 200, 201–2
Utopianism, 155, 159

Van Rensselaer, Martha, 126
Varieties of Man (Latham), 107
Vassar College, 70–71
Vestiges of the Natural History of Creation (Chambers), 102
Victorian Anthropology (Stocking), 106–7
Victorian era, 11, 14, 98, 106, 192; and Roberts, 30, 39*n1;* astronomy during, 61–75; and Gamble, 147–63; and Carson, 207; and Kingsley, 222, 223
Virtual witnessing, 50

Waiser, W. A., 82
Wake, C. Staniland, 155
Wakefield, Priscilla, 8, 11, 46, 118, 121–22
Walking with the Great Apes (Montgomery), 243
Wallace, Alfred Russel, 12
Walton, Izaak, 81
Ward, Lester, 161*n7*
Ward, Hon., Mrs., 174–75
"Water Beaux" (Meredith), 110–12
Ways of the Six-Footed (Comstock), 122–24, 134, 136
Wesley, W. H., 71
West African Studies (Kingsley), 225, 227
Westwood, J. O., 109
White, Gilbert, 81, 93
Whitman, Walt, 196, 262
Williams, Terry Tempest, 209–10
Williamson, Henry, 200
Winners in Life's Race (Buckley), 16, 165–68, 172
Woman in the Mists (Mowat), 250
Woman Today, 184–87
Women and Economics (Gilman), 158
Women's rights, 122, 159
Wood, J. G., 16
Woods Hole Marine Biological Laboratory, 199
Woolf, Virginia, 255
Working class, 46, 47, 218
World War II, 199
Writing Biology (Myers), 11

Young Emigrants, The (Traill), 83–84
"Young Gentleman and Lady's Philosophy" series, 6
Young Naturalist's Journey (Loudon), 12
Youth's Companion, 118

Zeller, Suzanne, 82
Zihlman, Adrienne, 162*n12*
Zoology, 84, 120, 199
Zoology of the Voyage of the Beagle (Bell), 101, 108–9
Zornlin, Rosina M., 13–14

SCIENCE AND LITERATURE
A series edited by George Levine

One Culture: Essays in Science and Literature
Edited by George Levine

In Pursuit of a Scientific Culture: Science, Art, and Society in the Victorian Age
Peter Allan Dale

Sexual Visions: Images of Gender in Science and Medicine between the Eighteenth and Twentieth Centuries
Ludmilla Jordanova

Writing Biology: Texts in the Social Construction of Scientific Knowledge
Greg Myers

Gaston Bachelard, Subversive Humanist: Texts and Readings
Mary McAllester Jones

Realism and Representation: Essays on the Problem of Realism in Relation to Science, Literature, and Culture
Edited by George Levine

Science in the New Age: The Paranormal, Its Defenders and Debunkers, and American Culture
David J. Hess

Fact and Feeling: Baconian Science and the Nineteenth-Century Literary Imagination
Jonathan Smith

The Word of God and the Languages of Man: Interpreting Nature in Early Modern Science and Medicine. Volume 1: Ficino to Descartes
James J. Bono

Seeing New Worlds: Henry David Thoreau and Nineteenth-Century Natural Science
Laura Dassow Walls

Natural Eloquence: Women Reinscribe Science
Edited by Barbara T. Gates and Ann B. Shteir